T0183526

Springer Textbooks in Earth Sciences, Geography and Environment

The Springer Textbooks series publishes a broad portfolio of textbooks on Earth Sciences, Geography and Environmental Science. Springer textbooks provide comprehensive introductions as well as in-depth knowledge for advanced studies. A clear, reader-friendly layout and features such as end-of-chapter summaries, work examples, exercises, and glossaries help the reader to access the subject. Springer textbooks are essential for students, researchers and applied scientists.

More information about this series at ▶ http://www.springer.com/series/15201

Paul Alexandre

Practical Geochemistry

 Springer

Paul Alexandre
Brandon University
Brandon, MB, Canada

ISSN 2510-1307   ISSN 2510-1315 (electronic)
Springer Textbooks in Earth Sciences, Geography and Environment
ISBN 978-3-030-72455-9   ISBN 978-3-030-72453-5   (eBook)
https://doi.org/10.1007/978-3-030-72453-5

© Springer Nature Switzerland AG 2021
This work is subject to copyright. All rights are reserved by the Publisher, whether the whole or part of the material is concerned, specifically the rights of translation, reprinting, reuse of illustrations, recitation, broadcasting, reproduction on microfilms or in any other physical way, and transmission or information storage and retrieval, electronic adaptation, computer software, or by similar or dissimilar methodology now known or hereafter developed.
The use of general descriptive names, registered names, trademarks, service marks, etc. in this publication does not imply, even in the absence of a specific statement, that such names are exempt from the relevant protective laws and regulations and therefore free for general use.
The publisher, the authors and the editors are safe to assume that the advice and information in this book are believed to be true and accurate at the date of publication. Neither the publisher nor the authors or the editors give a warranty, expressed or implied, with respect to the material contained herein or for any errors or omissions that may have been made. The publisher remains neutral with regard to jurisdictional claims in published maps and institutional affiliations.

This Springer imprint is published by the registered company Springer Nature Switzerland AG
The registered company address is: Gewerbestrasse 11, 6330 Cham, Switzerland

To my family

Preface

This textbook had an unusual inception. I had just started writing another one, on isotopes in the natural sciences, and was discussing it with a colleague, who surprised me and startled me. He commented that, in substance, a book about isotopes is all very nice, but that I should write a general introductory geochemistry book. I was taken aback: there are plenty of these out there already! From Brownlow [1] to Albarede [2], Misra [3], and White [4], to name just a few, there are several excellent general introductory geochemistry textbooks available, widely used by university students around the world. And once we include more specialized books, for instance about isotopes or geochronology, then we can fill not only shelves, but whole library sections with them! And really, I pointed out to my colleague that they are mostly very good, thorough, and complete textbooks, so the topic is well covered. Ah yes, he retorted, so they are, but they are not good for me, not at all. They are, he explained, much too theoretical and not practical at all. They are very good for those students who will likely later specialize in geochemistry; they are an excellent first step for them. They are written by geochemists for future geochemists. However, my colleague pointed out that they are not practical: they have plenty of theory (thermodynamics, bonding, equilibria, reactions, and so on), which often takes a third of the book and sometimes more than a half (as in Misra [3]). But the majority of the practising geologists do not need this theory in their everyday work, nor do the students just beginning to learn about geochemistry. Some very practical concepts (for instance, sampling, analytical techniques, data treatment, isocons, Pearce element ratios, spatial statistics, and many many others) are not mentioned at all in any of the general geochemistry books listed above. The same applies to mineral exploration, which is not mentioned at all in any geochemistry books; the beginner geochemist who wishes to use geochemistry for exploration must go directly to and struggle with some heavy-duty books on the topic, while they should be learning the basics. No doubt, my colleague concluded, that geochemistry has a reputation of being esoteric and abstract! And he left it at that.

My mind of an experienced geochemist rebelled at these assertions. Not true, I exclaimed! These are excellent books, thoughtful, detailed, complete, and up to date; I have them all and use them often. But the seed of doubt had been planted. The more I protested, the more I felt that maybe, just maybe, my colleague could be right. I scrutinized the books I have, on both general geochemistry and more specialized ones, and reluctantly came to the realization that yes, they are not practical. Somewhere in me, the notion that I should write this book started to grow as I was more and more convinced that yes, the world does need another geochemistry book: a practical and straightforward one.

So, that is how this book came about, and this is how its topic was defined. The entire purpose of this book is to provide a simple, applied, and down-to-earth introduction to geochemistry for students who are introduced to the subject for the first time, but also for practising geologists who don't need the theory, but rather some clear and simple practical pointers. After all, many students and practitioners will only need a fundamental and general, but also very practical and applied, understanding of geochemistry. They hardly get this from the geochemistry books currently available.

This book covers the basic and most relevant principles of geochemistry, explaining them whenever possible in plain terms. It avoids all underlying fundamental thermodynamic or physical chemistry principles. It also assumes that the reader has a solid senior high school or introductory university level of chemistry on topics such as reactions, pH and Eh, thermodynamics, and the like. It focuses on the geochemical behaviour of the elements (with a simple and short excursion into isotopes), based on some of their relevant characteristics. The analytical part is kept to a minimum, but covers the most common techniques. More than a half of the book is dedicated to practical applications, such as rock classification and provenance, detection of geochemical variations, mineral exploration, and environmental geochemistry.

Most significantly, this book is written from the point of view of a geologist and not of a geochemist: the specialist geochemist often has a chemical and analytical point of view, which sometimes proves to be irrelevant or inadequate to a practising geologist or an undergraduate student. This book also provides plenty of real-world and specific geological examples that make the basic concepts easily understood and integrated by a geologist.

Let me take the time to recognize all those who helped me along the way and significantly helped me write this book. These include, in no particular order, my colleague who planted the seed of doubt and set me on the way to this project; my colleagues in the academic world, for all their support and help; and the Springer editorial board, for all the practical advice and help, but also for their support and understanding. Chief among all these people comes my family who has been very patient, supportive, and helpful, and to whom this book is dedicated. To all of these, my heartfelt and deepest thank you!

References

1. Brownlow AH (1979) Geochemistry. Prentice-Hall, Englewood Cliffs, NJ, p 498
2. Albarede F (2009) Geochemistry, an introduction, 2nd edn. Cambridge University Press, Cambridge
3. Misra KC (2012) Introduction to Geochemistry: Principles and Applications. Wiley-Blackwell, Oxford, p 452
4. White WM (2013) Geochemistry. Wiley-Blackwell, Oxford

Note to the Instructor

Dear instructor,

I do hope that this book helps you not only to effectively teach the fundamentals of geochemistry to undergraduate students, but also to foster their curiosity in the subject. The book was purposefully written in a straightforward manner, with little hardcore theory, but with a very practical approach. It is intended as the first geochemistry book any student will encounter, and the idea is to demystify geochemistry, which has long suffered from a reputation of being esoteric and abstract. You and I know that this is not the case that on the contrary geochemistry is very logical and clear, and that its fundamental principles are purely rational and reasonable. It is our task to help the students understand that.

The book is organized in two major parts, some theory first (the elements and analytics) and then the major applications (lithogeochemistry, exploration, and environment). This gives you the opportunity to use the book as you wish, not necessarily in this order, and to use other sources and examples as needed. The review questions at the end of each chapter are developed as suggestions and can be modified as needed: you are the instructor, so please feel free to use this book as is best for you and for your students. Finally, there are plenty of books and references that you can use to extend your teaching as far as you feel confident. Fare well!

Contents

1	**The Elements**	1
1.1	Introduction	2
1.2	Formation of the Elements, or Nucleosynthesis	2
1.3	Characteristics of the Elements	2
1.4	Geochemical Classifications of the Elements	4
1.4.1	Classification by Abundance	4
1.4.2	Classification by the Presence in a Reservoir	5
1.5	Isotopes	6
1.5.1	Definition of Isotope	6
1.5.2	Isotopic Fractionation	7
1.5.3	Radioactive Disintegration	9
1.6	Behaviour of Elements in the Geological Environment	9
1.6.1	Mobility	12
1.6.2	Compatibility	13
1.6.3	Substitutions	13
1.7	Summary	15
References		15

2	**Analytical Methodology and Data Treatment**	17
2.1	Sampling	18
2.1.1	Purpose of Sampling	18
2.1.2	Representativity of Sampling	18
2.1.3	Sampling Scheme	18
2.1.4	Quality Control	18
2.1.5	Size of Sample	19
2.1.6	Types of Samples	19
2.1.7	Taking Care of Samples	20
2.2	Sample Treatment	21
2.2.1	Sample Preparation	21
2.2.2	Dissolution	22
2.3	Most Common Analytical Methods	22
2.3.1	Atomic Spectroscopy	22
2.3.2	Quadrupole Mass Spectrometer	24
2.3.3	Portable XRF	26
2.4	Measuring Isotopes	26
2.4.1	Mass Spectrometer	26
2.5	Microbeam Analyses	27
2.5.1	Electron Microprobe	27
2.5.2	Laser Ablation	28
2.5.3	Secondary Ion Mass Spectrometry	28
2.6	Data Treatment and Presentation	28
2.6.1	Purpose of Data Treatment	28
2.6.2	Single Variable	29
2.6.3	Anomaly	30
2.6.4	Two Variables	31
2.6.5	Multi-variate Analysis	33
2.7	Summary	33
References		34

3	**Lithogeochemistry**	35
3.1	Introduction: Definition and Applications	36
3.2	Whole Rock Classifications and Variability	36

3.2.1	Classification Diagrams	36
3.2.2	Classification Diagrams for Igneous Rocks	38
3.2.3	Classification Diagrams for Sedimentary Rocks	38
3.3	**Variation Diagrams**	39
3.4	**Discrimination Diagrams: Tectonic Setting and Mantle Reservoirs**	40
3.4.1	Tectonic Provenance	40
3.4.2	Mantle Heterogeneity	41
3.4.3	Magma Differentiation, Source of Sediments, and Degree of Alteration	45
3.4.4	Magma Differentiation Through Partial Melting and Fractional Crystallization	45
3.4.5	Rare Earth Elements as Proxy for Magma Differentiation	47
3.4.6	Rare Earth Elements in Sedimentary Rocks	48
3.4.7	Quantification of Hydrothermal Alteration	50
3.4.8	Pearce Element Ratios	51
3.4.9	Mixing Calculations	53
3.5	**Geochronology**	54
3.5.1	General Principles of Geochronology	54
3.5.2	Radiogenic Product Retention and Closure Temperature	55
3.5.3	Correction for Common Lead	56
3.5.4	Concordia Diagram	57
3.6	**Summary**	58
References		59
4	**Geochemical Exploration**	61
4.1	**Introduction and General Principles**	62
4.1.1	Definition and Purpose of Geochemical Exploration	62
4.1.2	The Concept of Source-Transport-Trap	62
4.1.3	Anomaly	63
4.2	**Distribution and Dispersion of Elements and Formation of Surface Anomalies**	63
4.2.1	Geochemical Behaviour of the Elements	63
4.2.2	Pathfinder Elements	63
4.2.3	The Mobilization of Elements	63
4.2.4	Formation of a Surface Anomaly	66
4.3	**Understanding the Formation and Evolution of a Mineral Deposit**	68
4.3.1	Fluids Involved in Deposit Formation	68
4.3.2	Physical Conditions of Deposit Formation	69
4.3.3	Chemical Conditions of Deposit Formation	70
4.3.4	Processes Involved in Deposit Formation	71
4.3.5	Geochronology in Metallogeny	72
4.4	**Detection of Surface Anomalies**	72
4.4.1	Quantification of Alteration	72
4.4.2	Soil Sampling	73
4.4.3	Sampling of Stream Sediments	76
4.4.4	Sampling of Vegetation	76
4.4.5	Sampling of Surface Waters	76
4.4.6	Survey Design	76
4.5	**Isotopes in Exploration**	77
4.6	**Data Interpretation: Spatial Geostatistics**	78
4.6.1	Summary Statistics	78
4.6.2	Spatial Data Visualization	79
4.6.3	Kriging	81
4.7	**Summary**	82
References		82
5	**Environmental Geochemistry**	85
5.1	**The Significance of Environmental Geochemistry**	86
5.1.1	The Anthroposphere	86

5.1.2   Applications of Environmental Geochemistry 86
5.2   **Past Climate Change** 87
5.2.1   The Historic Compilation of Guy Callendar 87
5.2.2   Climate Change Causes 88
5.2.3   Records of Paleo-Temperatures.. 88
5.2.4   Isotopes as Paleo-Temperature Proxies 89
5.3   **Tracing Industrial and Agricultural Pollution**....................................... 90
5.3.1   General Principles of Pollution Tracing... 90
5.3.2   Lead as Industrial Pollution Tracer 90
5.3.3   Nitrogen and Water Pollution 92
5.4   **Environmental Biogeochemistry and Remediation** 94
5.4.1   Definition of Biogeochemistry.. 94
5.4.2   Environmental Biogeochemistry and Remediation.. 94
5.4.3   Adsorption and Remediation... 96
5.5   **Summary** 98
References.. 98

Supplementary Information 101
Epilogue.. 102
Glossary 105
Index 111

The Elements

Contents

1.1   Introduction – 2

1.2   Formation of the Elements, or Nucleosynthesis – 2

1.3   Characteristics of the Elements – 2

1.4   Geochemical Classifications of the Elements – 4
1.4.1   Classification by Abundance – 4
1.4.2   Classification by the Presence in a Reservoir – 5

1.5   Isotopes – 6
1.5.1   Definition of Isotope – 6
1.5.2   Isotopic Fractionation – 7
1.5.3   Radioactive Disintegration – 9

1.6   Behaviour of Elements in the Geological Environment – 9
1.6.1   Mobility – 12
1.6.2   Compatibility – 13
1.6.3   Substitutions – 13

1.7   Summary – 15

        References – 15

Electronic supplementary material The online version of this chapter (▶ https://doi.org/10.1007/978-3-030-72453-5_1) contains supplementary material, which is available to authorized users.

© Springer Nature Switzerland AG 2021
P. Alexandre, *Practical Geochemistry*,
Springer Textbooks in Earth Sciences, Geography and Environment,
https://doi.org/10.1007/978-3-030-72453-5_1

1

1.1   Introduction

The chemical elements are the essential building blocks of any physical material, including rocks and minerals. It is therefore important that, before going any further, we familiarize ourselves with the chemical elements and find out who they are (including their isotopes) and how, when, and where they formed. We will also pay particular attention to their characteristics—in particular those relevant to geological processes—and therefore to their geochemical behaviour.

1.2   Formation of the Elements, or Nucleosynthesis

Let us first have a very brief overview of how and where the chemical elements and their isotopes were and continue to be formed. Simplistically, there are four main settings where nuclides were and are produced:

1. *During the Big Bang*, between 1 s and 3 min after the beginning of the universe, when temperatures became sufficiently low (around 10^9 K) to allow the formation of nuclear particles and complex nuclei. Firstly, all of H (consisting of one proton) was formed at that point, followed by He, and minute amounts of Li, Be, and possibly B. Temperature and, to a lesser extent, pressure were too high for any other nuclide to be stable.

2. 2. *Stellar nucleosynthesis*, which occurred in a star from the main sequence (where it resided during most of its life) and consisted of three episodes of nuclear fusion. Here, nuclides combined to form heavier and more stable isotopes, releasing energy in the process.
    - Episode 1, proton addition. Small amounts of H were produced, together with some He, by the major and very slow proton–proton reaction (PPI). This was followed by the minor PPII and PPIII processes, responsible for the formation of elements up to C. This stage lasts a few to several millions of years.
    - Episode 2, the C–N–O cycle. A series of proton additions, combined with γ and $\beta+$ decays, formed all C, O, and N isotopes. This stage is faster: it lasts thousands to hundreds of years.
    - Episode 3, carbon, oxygen, and nitrogen burning. By using C, O, and N as the main "fuel" and combining them with lighter elements or individual particles, all elements up to Fe were formed. Things are much faster now: these stages last years to days.
    A major characteristic of stellar nucleosynthesis is that it occurs in conditions of mechanical equilibrium, resulting in the formation of predominantly stable isotopes; this is why the light isotopes are predominantly stable, with very few exceptions. The most stable element, Fe, is also the last to be produced in a star from the main sequence.
    Another characteristic of stellar nucleosynthesis is the constant release of energy due to nuclear fusion

and isotopes occupying lower energy states. A large proportion of this energy will be emitted outside of the star, but some will cumulate, leading eventually to a brutal departure from the state of equilibrium and the formation of Super-Nova.

3. *Supernova*: explosive nucleosynthesis. We are no longer in equilibrium, and a rapid addition of neutrons (the s-process, occurring in a semblance of equilibrium) is combined with β-decay (the *r*-process, in disequilibrium). These two processes form all the remaining naturally occurring elements, all the way to U. Significantly, the s-process will tend to form stable isotopes (the heaviest of which is ^{209}Bi), while the *r*-process will create mostly radioactive isotopes.

4. There is another setting where elements are formed: *outer space*, where *spallation* occurs. The process consists of nuclei colliding with a high-energy cosmic ray (consisting mainly of fast protons) and producing small amounts of light elements, such as H, He, Li, Be, and B. These isotopes are actually consumed at temperatures involving hydrogen burning in a star and are unstable at Big Bang conditions, which is why spallation is the only process that produces them. However, with collision probability in outer space being exceedingly low, we end up with very low amounts of these isotopes (◘ Fig. 1.1).

1.3   Characteristics of the Elements

There are many characteristics of the chemical elements that we can consider and study, but we will limit ourselves only to those characteristics that are the most relevant to the elements' behaviour in the geological environment. Specifically, these are atomic number, size, mass, charge, and electronegativity.

Atomic number. Each chemical element is defined by a unique number, corresponding to the number of protons in its nucleus, from 1 for H to 92 for U (elements beyond uranium, with atomic numbers 93 to 109, are very short-lived and are present in the geological environment only in exceedingly small amounts—we will not consider them here).

Size. The ionic radii of the elements vary as function of their state (bound or free), their coordination number (or how many ions are in immediate vicinity in the crystalline structure of a mineral), and oxidation, or valance, state. As a result, we will report them according to these factors (and refer to them as *effective ionic radii*), but will also consider the average ionic radius for each element. Those vary from "very small" for H to approximately 1.8 Å for Cs (pronounced *angstrom*; $1\,\text{Å} = 10^{-10}$ m, or 0.1 nm). There is a general trend of elements becoming larger in the higher periods of the periodic table.

Ionic charge. The charge of an element is determined by the number of electrons in its outermost, valance elec-

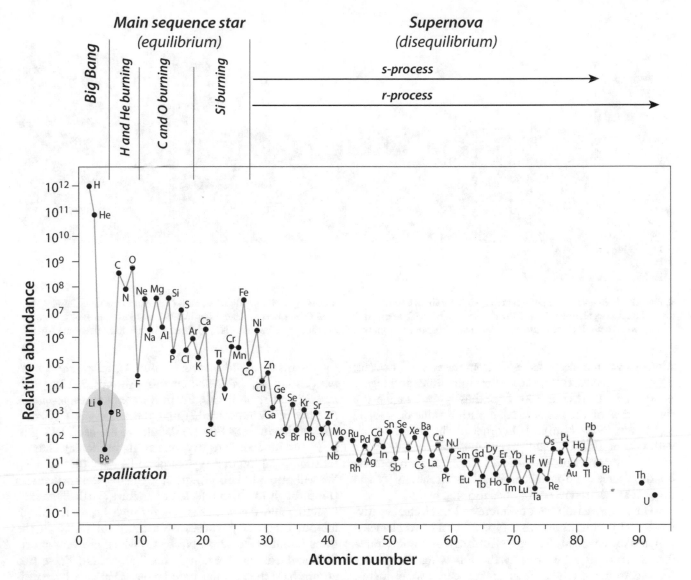

◘ Fig. 1.1 Relative abundance of the chemical elements in the solar system, taken as proxy for the universe (from [1]). A few regular features are evident: H and He and by far the most abundant, then we have a gradual decrease with a peak at Fe, and a sharp drop for Li, Be, and B. This distribution is explained by the formation of different elements at different times, in different places, by different processes, and under different conditions, as described in the text

tron shell. That shell has its full complement of 8 electrons only for the noble gases , which means that they do not need other electrons and thus do not react with each other or with any other element under normal geological conditions (hence the name *noble*). For all other elements, electrons from the valance shell will be exchanged and (ionic or covalent) bonds will be formed resulting in complete valence shells. The number of electrons an element has will thus differ from the number of protons and the difference defines the ionic charge of an element.

The ionic charge varies from −3 to +7 and can be deduced from the position of the element in the periodic table. All elements in Group 1 have a charge of +1, those of Group 2 have a charge of +2; the transition metals, the semi-metals, and the lanthanides will have charges of +2 to +5 (mostly +3 and rarely +6 and +7); the actinides will have charges of +3 to +6; and the non-met-

als will have negative charges, from −1 to −3 (the noble gases have charges of 0, as noted earlier). In general, metals tend to be cations (with positive charges), whereas non-metals tend to be anions (with negative charges); there are very few exceptions to the trend.

Atomic mass. The mass of an element is given by the number of protons plus the number of neutrons in its nucleus, each of which has a mass of 1 atomic mass unit. (The mass of the electrons being exceedingly low, we will assume it to be 0.) The mass of elements will vary from 1 for H to approximately 238 for U. As many elements have a varying number of neutrons, as we'll see later, their mass will not be an integer number.

Electronegativity. This is, simplistically, the tendency of an atom to attract a shared pair of electrons (or electron density) towards itself. An atom's electronegativity is affected by both its atomic number and the distance at

1 H 2.1																	
3 Li 1.0	4 Be 1.5											5 B 2.0	6 C 2.5	7 N 3.0	8 O 3.5	9 F 4.0	
11 Na 1.0	12 Mg 1.2											13 Al 1.5	14 Si 1.8	15 P 2.1	16 S 2.5	17 Cl 3.0	
19 K 0.9	20 Ca 1.0	21 Sc 1.3	22 Ti 1.4	23 V 1.5	24 Cr 1.6	25 Mn 1.6	26 Fe 1.7	27 Co 1.7	28 Ni 1.8	29 Cu 1.8	30 Zn 1.6	31 Ga 1.7	32 Ge 1.9	33 As 2.1	34 Se 2.4	35 Br 2.8	
37 Rb 0.9	38 Sr 1.0	39 Y 1.2	40 Zr 1.3	41 Nb 1.5	42 Mo 1.7	43 Tc 1.7	44 Ru 1.8	45 Rh 1.8	46 Pd 1.8	47 Ag 1.6	48 Cd 1.6	49 In 1.6	50 Sn 1.8	51 Sb 1.9	52 Te 2.1	53 I 2.5	
55 Cs 0.8	56 Ba 1.0	57-71 La-Lu 1.1	72 Hf 1.3	73 Ta 1.4	74 W 1.5	75 Re 1.7	76 Os 1.9	77 Ir 1.9	78 Pt 1.8	79 Au 1.9	80 Hg 1.7	81 Ti 1.6	82 Pb 1.7	83 Bi 1.8	84 Po 1.9	85 At 2.1	
87 Fr 0.8	88 Ra 1.0	89-103 Ac-Lr 1.1															

■ **Fig. 1.2**   Elements in the periodic table, with Pauling's electronegativity values [6], which are dimensionless (The American Linus Pauling, who defined electronegativity, achieved the rare exploit of being awarded the Nobel Prize twice: in Chemistry, in 1954, and Peace, in 1962). Note that noble gases are not included here: they are unreactive under normal conditions and therefore there is no defined value for their electronegativity

which its valence electrons reside from the charged nucleus; it typically decreases with lower group number and higher period number (■ Fig. 1.2). It is thus highest in the top right corner of the periodic table, with a value of 4 for F, and lowest in the bottom left corner of the periodic table, with 0.7 for Fr. In general terms, metals have lower electronegativity and thus act as electron donors when forming bonds, whereas non-metals have high electronegativity and thus act as electron receivers when forming bonds.

The main reason we are interested in electronegativity is that it conditions what type of bond two elements will make, covalent (shared electrons) or ionic (transferred electrons). The rule is the following: the higher the electronegativity difference between two elements, the higher the probability that these elements will form an ionic bond (and the lower the probability they will form a covalent bond). For example, Na and Cl will very likely form an ionic bond in halite, as their electronegativity difference is high (2.0), whereas C and O in CO_2 will likely form a covalent bond. A molecule of the same element (e.g., O_2) will thus have uniquely covalent bond between the two atoms.

1.4   Geochemical Classifications of the Elements

There are many ways to group the chemical elements in separate categories. We can use a particular characteristic (e.g., size, weight), or their place in the periodic table of the elements, or their belonging in a particular reservoir, or their behaviour. The simplest way would be to separate the chemical elements into *metals*, *non-metals*, and *gases*, which is neither sufficiently detailed nor sufficiently specific, but relies on broadly defined common characteristics and behaviour. We can then subdivide

the metals in *alkali metals* (Group 1), *alkali-earth metals* (Group 2), *transition metals* (Groups 3–12), *other metals*, and *semi-metals*. To that we will add *non-metals* (mostly in Groups 15–17) and the *noble gases* (Group 18), as shown in ■ Fig. 1.3. While we are at it, let's give hydrogen its own category, due to its particular characteristics, resulting in a typical periodic table (■ Fig. 1.3). We will also add two groups of elements, the *lanthanides* (La through Lu; also referred to as Rare Earth Elements, or REE) and the *actinides* (Ac through Lr, even though for our practical purposes we are not interested in the trans-uranium elements, Np to Lr, which are very short-lived and exist in exceedingly small amounts). These two groups have the peculiarity to be placed in only two positions in the periodic table of the elements, one for the lanthanides and one for the actinides, and thus have very similar chemical characteristics and behaviour.

This classification of the elements is very common, and we will rely on it and use it extensively; it is the initial part of our geochemical vocabulary. However, it is not specific to geology and is not particularly helpful from the point of view of the geological environment and processes. For our purposes, we have to consider some other factors, relevant to geology, namely the elements' abundance on Earth and where they are most likely to be found.

1.4.1   Classification by Abundance

For any practical purpose, this is the most important classification. We will separate the elements in the groups of *major*, *minor*, and *trace* elements (■ Fig. 1.4), even though in common practice only the groups of *major* and *trace* elements are used:

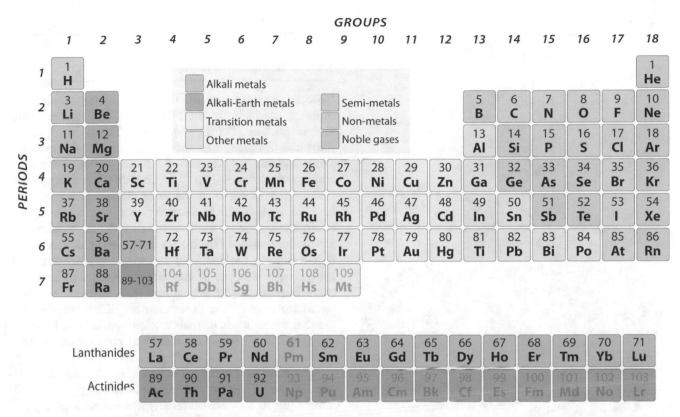

Fig. 1.3 A general classification of the chemical elements, presented in the form of the Periodic Table of the elements. Elements with atomic numbers above 92, shown in grey, are very short-lived and exist in exceedingly small amounts

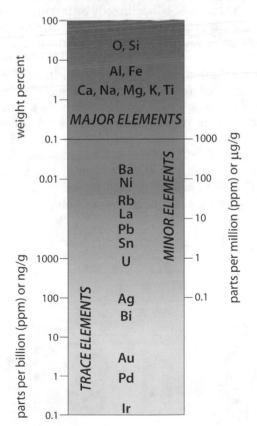

Fig. 1.4 Classification of the chemical elements based on their abundance in Earth's continental crust, with approximate position of a few selected elements. This classification may be further simplified to simply major and trace elements. Data from CRC Handbook of Chemistry and Physics 2016–2017 [2]

— Major elements (Si, Al, Mg, Fe, Ca, Na, K, and Ti): these are the main constituents of a rock and are very likely to be present in concentrations above 0.1 wt%. Their concentrations are reported as weight percent of oxides (wt%). The oxides of the major elements account for approximately 99.7 wt% of the elements present in rocks on the continents; because of that we should always remember who they are. Please note that P may also be considered member of this group.

— Minor elements (e.g., Ba, Ni, Rb, La, Pb, Sn): any element at around 0.1 wt% is placed here and also reported as wt% oxide.

— Trace elements (e.g., Au, Pd, Ir): these are trace constituent of rocks and minerals and are reported in parts per million (ppm) or parts per billion (ppb).

1.4.2 Classification by the Presence in a Reservoir

Because of their specific characteristics and behaviour, the elements tend to concentrate in specific Earth reservoirs (**Fig. 1.5**). Many of them belong to more than one reservoir, of which there are four:

— Lithophile, or belonging to the rock domain (i.e., Earth's mantle and crust). The main ones are Si, Al, Mg, Na, Ca, K, and Fe, with minor ones such as Li, Be, B, the Rare Earths, U, Th, etc.

— Siderophile, or belonging to Earth's ore. These are Fe and Ni, with very few other ones.

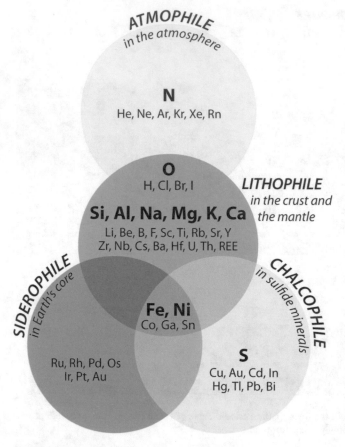

ATMOPHILE
in the atmosphere

N
He, Ne, Ar, Kr, Xe, Rn

O
H, Cl, Br, I

Si, Al, Na, Mg, K, Ca
Li, Be, B, F, Sc, Ti, Rb, Sr, Y
Zr, Nb, Cs, Ba, Hf, U, Th, REE

LITHOPHILE
in the crust and the mantle

SIDEROPHILE
in Earth's core

CHALCOPHILE
in sulfide minerals

Fe, Ni
Co, Ga, Sn

Ru, Rh, Pd, Os
Ir, Pt, Au

S
Cu, Au, Cd, In
Hg, Tl, Pb, Bi

◘ Fig. 1.5 Classification of the chemical elements based on which major reservoir they are preferentially found in. The most abundant elements in the geological realm are given in bold, large letters. Note that an element can be shared between 2 reservoirs (O) or even three (Fe, Ni)

— Atmophile, or belonging to the atmosphere. These are mostly O, N, and the noble gases.
— Chalcophile, or belonging in sulphide minerals. The main ones are S, Fe, and Ni, with also Cu, Ag, Cd, Zn, In, Pb, Bi, etc.

Again, this is an important classification and, as such, makes part of our vocabulary of geochemists; this is why we should be aware at any point which reservoir any element would belong to preferentially.

1.5   Isotopes

1.5.1   Definition of Isotope

Isotopes are variants of the same chemical element differing uniquely by the number of neutrons in their atomic nucleus (◘ Fig. 1.6). The different isotopes of the same chemical element will be found in the exact same position in the periodic table (hence the name, which means "*same place*" in Greek). That means that the different isotopes of the same chemical element will have the same chemical characteristics, such as electronic configuration, ionic radius, valence states, electronegativity, and so on. Given that it is these characteristics that condition the chemical behaviour of the elements, the different isotopes of the same element will have the exact same chemical behaviour. In a nutshell, the different isotopes of the same element differ only by the number of neutrons in their nucleus—and thus by their mass—but have the same chemical characteristics and behaviour.

There are two types of isotopes, stable and radioactive. The stable ones are not changed in any way and will remain just as they are under any normal geological conditions, whereas the radioactive ones—as they are unstable—disintegrate to stable isotopes, at a known rate and by known processes, either directly or through a series of disintegrations. (The stable isotopes that are the product of radioactive disintegration are called radiogenic isotopes.)

By convention, the isotopic composition of a sample is expressed relative to that of a specific standard, which is different for different elements. In all cases, we are interested in the ratio of the minor, heavier isotope versus the lighter, major one, for instance $^2H/^1H$, or $^{18}O/^{16}O$, or $^{13}C/^{12}C$, and so on. The ratios are compared to those of a standard and are expressed by the δ-notation ("delta notation"; O'Neil 5), using the following equation, for the example of oxygen isotopes:

$$^{18}O_{SAMPLE} = \left[\frac{\left(^{18}O/_{16}O\right)_{SAMPLE} - \left(^{18}O/_{16}O\right)_{STANDARD}}{\left(^{18}O/_{16}O\right)_{STANDARD}} \right] \times 10^3, ‰$$

◘ Fig. 1.6 Schematic representation of the nuclei of the three main isotopes of carbon (^{12}C, ^{13}C, and ^{14}C), with the number of protons and number of neutrons. They all contain six protons, which defines their chemical behaviour, but differ in the number of neutrons (6, 7, and 8, respectively), resulting in different masses

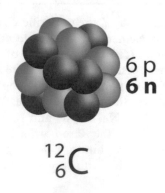

6 p
6 n

$^{12}_6C$

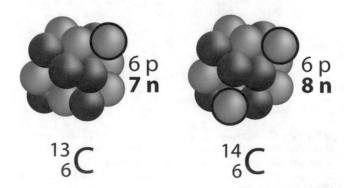

6 p
7 n

$^{13}_6C$

6 p
8 n

$^{14}_6C$

The practical meaning of this notation is that a high δ value indicates a higher proportion of the heavier isotope relative to a standard. A positive δ value indicates that the sample is enriched in the heavy isotope relative to the standard, whereas a negative δ value indicates that the heavy isotope is depleted in the sample relative to the standard. When a sample has a δ value of 0, it indicates that the sample has the same proportion of the heavy isotope as the standard.

As mentioned earlier, the different isotopes of the same element have the exact same chemical characteristic and thus behaviour. However, they differ in mass and that difference, even though sometimes small, will cause different isotopes to partition, or segregate, between different compounds or phases during a range of geological processes. This phenomenon is called *isotopic fractionation.*

1.5.2 Isotopic Fractionation

Geological processes are either very slow, occur at high temperatures, or both, which means that *isotopic equilibrium* is very likely to be achieved between two phases (e.g., minerals and fluids) involved in a reaction during these processes. The resulting *equilibrium isotopic fractionation* between the two phases is due uniquely to the differences in bonding energy between isotopes, itself due to differences in their mass. As a simple example, the three different H_2 molecules, $^1H_2$, $^1H^2H$, and $^2H_2$, have masses of 1, 3, and 4, and dissociation energies (the reverse of bonding energy) of 103.2, 104, and 105.3 kcal/mole, respectively, at standard pressure and temperature. In another example, let us consider two water molecules made of ^1H or ^2H and of ^{16}O: $^1H_2^{16}O$ and $^2H_2^{16}O$. The mass difference between the two molecules is 11% (mass of 18 vs. mass of 20), resulting in measurable difference in density (0.998 and 1.105 g/cm^3, respectively), melting temperature (0 °C and 3.82 °C, respectively), and boiling temperature (100 °C and 101.4 °C, respectively). These are very significant differences, and ones that will lead to noticeable fractionation between liquid water and water vapour, every time evaporation or condensation occurs.

However, it should be noted that there are several important geological processes that are biologically mediated, for instance sulphate reduction or carbon incorporation into biomass. In this situation, isotope fractionation often does not occur under equilibrium conditions. We call this *kinetic* fractionation, which is the dominant fractionation when rapid and non-reversible (unidirectional) reactions occur (for instance during evaporation and diffusion), or in any situation when equilibrium has not been fully reached. The main cause for kinetic fractionation is the difference in diffusion rates between isotopes of the same elements (itself due to differences in activation energy between two isotopes

of different mass). Indeed, lighter compounds—those made of lighter isotopes—will diffuse faster than heavier ones, as diffusion rate is function of mass. The result of kinetic fractionation is that the products of a reaction will always be lighter (depleted in the heavy isotope) relative to the reactants. Kinetic fractionation is typically very large, often an order of magnitude larger than equilibrium fractionation (◨ Fig. 1.7).

When considering equilibrium fractionation (which is by far the most common in geological processes), we can say that heavier isotopes of the same element form stronger bonds than lighter isotopes because of their higher mass, and thus produce more stable molecules. One major consequence of this effect can be found in the general equilibrium fractionation rule stipulated by Jacob Bigeleisen in 1965: "*The heavy isotope goes preferentially to the chemical compound in which the element is bound most strongly.*" We can rephrase this to say that, under equilibrium fractionation conditions, heavy isotopes tend to concentrate in phases with stronger chemical bonds.

Another consequence of the dissociation energy variations is that the extent of fractionation (how much fractionation will occur in a given chemical or physical process) will depend directly on the temperature at which this process occurs. As demonstrated in ◨ Fig. 1.8, the extent of fractionation is always higher at lower temperature, and vice versa, an important and invariable rule.

In order to quantify the extent of fractionation between two phases or compounds in equilibrium, we use the so-called fractionation factor, α (alpha). It is defined as the ratio between the isotope ratios in two compounds at equilibrium:

$$\alpha = \frac{\left(^{18}O/_{16}O\right)_{\text{phase 1}}}{\left(^{18}O/_{16}O\right)_{\text{phase 2}}}$$

From a practical point of view, and considering the definition of δ, some mathematical transformations allow us to approximate the difference between the isotopic composition of two compounds to 1000 lnα:

$$\delta_A - \delta_B \sim \ln \alpha \times 1000$$

Crucially, and as explained above, this difference is always a function of temperature. As a result, if the isotopic compositions of two compounds at equilibrium are known, the temperature at which the reaction occurs can be found. On the other hand, if the isotopic composition of one of the two compounds is known, and the temperature of the equilibrium reaction is obtained by some independent method, it is possible to calculate the isotopic composition of the other compound.

For instance, let us consider quartz precipitating from water under equilibrium conditions. It has been established that the fractionation factor, 1000 lnα, is 5‰ at

1

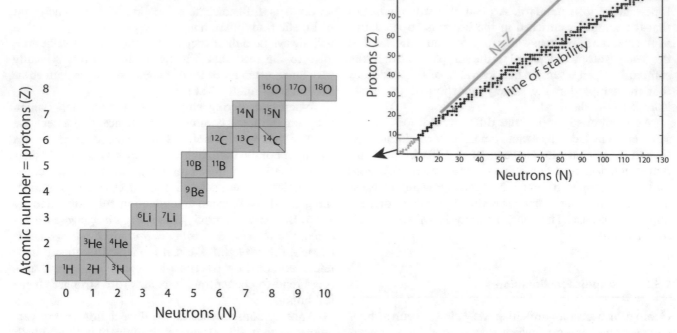

❏ **Fig. 1.7** Isotopes of the first eight elements (also known as *light isotopes*) in the chart of isotopes, where the nuclei are plotted by their number of protons and number of neutrons: each line corresponds to one chemical element and its various isotopes, each occupying one box and defined by their mass. Two isotopes are considered stable but are in fact radioactive: ^3H and ^{14}C are produced (by cosmic radiation in the upper atmosphere) at approximately the same rate at which they disintegrate, keeping their amount on Earth approximately constant. The line of stability is defined by the most "advantageous" combination of the number of protons and the number of neutrons that will produce the most stable nuclei. As more protons are added to make heavier elements, even more neutrons are required to counteract the repulsive longer-range electrostatic force (active between protons only): as a result, the line of stability will progressively go further away from the $N=Z$ line. Any isotope that is not on the line of stability, but above or below it, will be radioactive and will disintegrate to a radiogenic product (or "*daughter*" isotope), itself necessarily placed on the stability line

❏ **Fig. 1.8** Two different forces acting upon the two atoms of the O_2 molecule. The difference in potential energy between two oxygen molecules ($^{18}O_2$ and $^{16}O_2$) varies as function of temperature. After O'Neil [5], modified

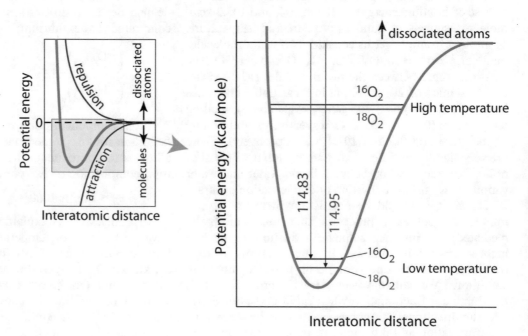

200 °C. Therefore, if the isotopic composition of quartz is −10‰, we can calculate that the water with which this quartz was in equilibrium with, had a δ of −15‰. If, alternatively, we have the isotopic composition of both quartz and water, we can in turn calculate the temperature at which this equilibrium reaction occurs.

1.5.3 Radioactive Disintegration

Equilibrium fractionation will affect both stable and radioactive elements, but is mostly studied for the former. The geological applications of the radioactive isotopes, on the other hand, are much more related to their disintegration, as we will see here.

Radioactive decay (or disintegration) occurs at a constant speed, which is known—with a high degree of precision—for each radioactive isotope. This means that if we know how much of the radioactive ("parent") isotope has disintegrated to produce the radiogenic ("daughter") isotope, and if we know the rate, or speed, of disintegration, we can calculate the age at which the parent isotope was incorporated into a mineral. This is the basis of *absolute*—or radiometric—geochronology (as opposed to *relative* geochronology based on stratigraphy). E. Rutherford, who first experimented with and basically invented geochronology, stated its fundamental principle: "The rate of decay of a radioactive nuclide is proportional to the number of atoms of that nuclide remaining at any time." Mathematically, this can be expressed with:

$$-\frac{dN}{dt} = \lambda N$$

where t is the time elapsed, λ is the disintegration constant (specific to each radioactive isotope and known with a certain level of precision), and N is the amount of radioactive parent remaining. After a series of mathematical transformations, we can solve this equation for the time t:

$$t = \frac{1}{\lambda}\ln\left(\frac{D^*}{N} + 1\right)$$

where D^* is the amount of daughter isotope coming directly from the disintegration of the parent and N is the amount of the remaining parent isotope. In other words, if we measure the amounts of N and D^* (expressed as number of atoms) in a mineral and know λ, we can calculate the age of that mineral. This general equation is applicable to any parent–daughter system. In practical terms, there are many factors and assumptions that strongly affect absolute geochronology and we will go into detail on this later in this book, in ▶ Chap. 3. On the other hand, there are other geological applications of radiogenic isotopes and we will discuss them Chapter 3.

1.6 Behaviour of Elements in the Geological Environment

The two most important aspects of the behaviour of chemical elements in the geological environment are their mobility, their compatibility, and their ability to substitute into the crystal structure of a mineral. These stem directly from the elements' characteristics, specifically ionic size and charge, and the prevailing physical and chemical conditions of the relevant processes.

A major consideration for both mobility and compatibility is the ionic potential (or charge density , or field strength) of an element, defined as the ratio of the ionic charge over ionic size (❏ Fig. 1.9). Ionic charge

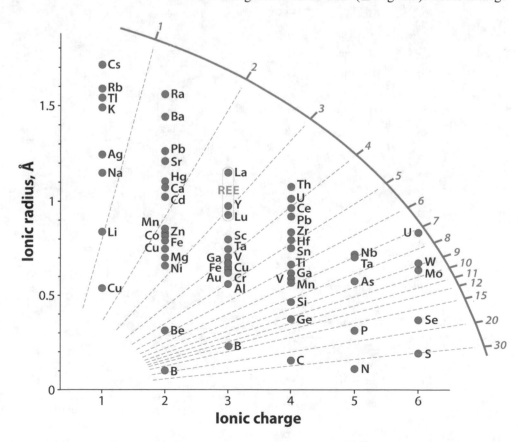

❏ **Fig. 1.9** Chemical elements plotted in the size versus charge diagram, using data from ❏ Table 1.1; [4, 7]. Their ionic potential (or density of charge) varies significantly between elements and directly conditions their behaviour in the geological environment

1

◨ Table 1.1   Major characteristics of the chemical elements, including a list of their major isotopes

Element			Ionic charge	Ionic radius (Å)			Electronegativity	Major isotopes
Atomic number	Symbol	Name	(Valance state)	cn4	cn6	cn8	Pauling scale	
1	H	Hydrogen	+1	*Extremely small*			2.1	^{1}H, ^{2}H, ^{3}H
2	He	Helium	0					^{3}He, ^{4}He
3	Li	Lithium	+1	0.68	0.82		1	^{6}Li, ^{7}Li
4	Be	Beryllium	+2	0.35			1.5	^{9}Be
5	B	Boron	+3	0.20			2	^{10}B, ^{11}B
6	C	Carbon	+2, +4		0.16		2.5	^{12}C, ^{13}C, ^{14}C
7	N	Nitrogen	−3 to +5		0.13		3	^{14}N, ^{15}N
8	O	Oxygen	−2		1.32	1.34	3.5	^{16}O, ^{17}O, ^{18}O
9	F	Fluorine	−1	1.23	1.25		4	^{19}F
10	Ne	Neon	0					^{20}Ne, ^{21}Ne, ^{22}Ne
11	Na	Sodium	+1	1.07	1.10	1.24	0.9	^{23}Na
12	Mg	Magnesium	+2	0.66	0.80	0.97	1.2	^{24}Mg, ^{25}Mg, ^{26}Mg
13	Al	Aluminium	+3	0.47	0.61		1.5	^{27}Al
14	Si	Silicon	+4	0.34	0.48		1.8	^{28}Si, ^{29}Si, ^{30}Si
15	P	Phosphorus	−3, +3, +5		0.44		2.1	^{31}P
16	S	Sulphur	+2, +4, +6	0.20	0.30		2.5	^{32}S, ^{33}S, ^{34}S, ^{36}S
17	Cl	Chlorine	−1	1.67	1.72	1.65	3	^{35}Cl, ^{37}Cl
18	Ar	Argon	0					^{36}Ar, ^{38}Ar, ^{40}Ar
19	K	Potassium	+1		1.46	1.56	0.8	^{39}K, ^{40}K
20	Ca	Calcium	+2		1.08	1.20	1	^{40}Ca, ^{42}Ca, ^{44}Ca, ^{48}Ca
21	Sc	Scandium	+3		0.83	0.95	1.3	^{45}Sc
22	Ti	Titanium	+3, +4		0.61		1.5	^{46}Ti, ^{47}Ti, ^{48}Ti, ^{49}Ti, ^{50}Ti
23	V	Vanadium	+2, +3, +4, +5		0.60		1.6	^{50}V, ^{51}V
24	Cr	Chromium	+2, +3, +6	0.42	0.62		1.6	^{50}Cr, ^{52}Cr, ^{53}Cr, ^{54}Cr
25	Mn	Manganese	+2, +4, +7	0.35	0.73		1.5	^{55}Mn
26	Fe	Iron	+2, +3	0.64	0.78		1.8	^{54}Fe, ^{56}Fe, ^{57}Fe, ^{58}Fe
27	Co	Cobalt	+2, +3	0.65	0.75		1.9	^{59}Co
28	Ni	Nickel	+2		0.69		1.9	^{58}Ni, ^{60}Ni, ^{61}Ni, ^{62}Ni, ^{64}Ni
29	Cu	Copper	+1, +2	0.70	0.81		1.9	^{63}Cu, ^{65}Cu
30	Zn	Zinc	+2	0.68	0.83	0.98	1.6	^{64}Zn, ^{66}Zn, ^{67}Zn, ^{68}Zn, ^{70}Zn
31	Ga	Gallium	+3	0.55	0.70		1.6	^{69}Ga, ^{71}Ga
32	Ge	Germanium	+2, +4	0.48	0.62		1.8	^{70}Ge, ^{72}Ge, ^{73}Ge, ^{74}Ge, ^{76}Ge
33	As	Arsenic	+3, +5	0.42	0.58		2	^{75}As
34	Se	Selenium	−2, +6	0.37	0.42		2.5	^{74}Se, ^{76}Se, ^{77}Se, ^{78}Se, ^{80}Se, ^{82}Se
35	Br	Bromine	−1, +7	1.88	1.96		2.8	^{79}Br, ^{81}Br
36	Kr	Krypton	0					^{78}Kr, ^{80}Kr, ^{82}Kr, ^{83}Kr, ^{84}Kr, ^{86}Kr
37	Rb	Rubidium	+1		1.57	1.68	0.8	^{85}Rb, ^{87}Rb
38	Sr	Strontium	+2		1.21	1.33	1	^{84}Sr, ^{86}Sr, ^{87}Sr, ^{88}Sr
39	Y	Yttrium	+3		0.98	1.10	1.2	^{89}Y
40	Zr	Zirconium	+4		0.80	0.86	1.4	^{90}Zr, ^{91}Zr, ^{92}Zr, ^{94}Zr, ^{96}Zr
41	Nb	Niobium	+3, +5		0.79		1.6	^{93}Nb

1.6 · Behaviour of Elements in the Geological Environment

◘ **Table 1.1** (continued)

Element			Ionic charge	Ionic radius (Å)			Electronegativity	Major isotopes
Atomic number	Symbol	Name	(Valance state)	cn4	cn6	cn8	Pauling scale	
42	Mo	Molybdenum	+3, +6	0.50	0.75		1.8	^{92}Mo, ^{94}Mo, ^{95}Mo, ^{96}Mo, ^{97}Mo, ^{98}Mo, ^{100}Mo
43	Tc	Technetium	+6		0.72			^{98}Tc
44	Ru	Ruthenium	+3, +4, +8		0.76		2.2	^{96}Ru, ^{98}Ru, ^{99}Ru, ^{100}Ru, ^{101}Ru, ^{102}Ru, ^{104}Ru
45	Rh	Rhodium	+4		0.73		2.2	^{103}Rh
46	Pd	Palladium	+2, +4	0.72	0.84		2.2	^{102}Pd, ^{104}Pd, ^{105}Pd, ^{106}Pd, ^{108}Pd, ^{110}Pd
47	Ag	Silver	+1	1.10	1.23		1.9	^{107}Ag, ^{109}Ag
48	Cd	Cadmium	+2	0.88	1.03	1.15	1.7	^{106}Cd, ^{108}Cd, ^{110}Cd, ^{111}Cd, ^{112}Cd, ^{113}Cd, ^{114}Cd, ^{116}Cd
49	In	Indium	+3		0.88	1.00	1.7	^{113}In, ^{115}In
50	Sn	Tin	+2, +4		0.93	1.30	1.8	^{112}Sn, ^{116}Sn, ^{117}Sn, ^{118}Sn, ^{119}Sn, ^{120}Sn, ^{122}Sn, ^{124}Sn
51	Sb	Antimony	+3, +5	0.85	0.76		1.9	^{121}Sb, ^{123}Sh
52	Te	Tellurium	+4, +6		0.70		2.1	^{120}Te, ^{122}Te, ^{123}Te, ^{124}Te, ^{125}Te, ^{126}Te, ^{128}Te, ^{130}Te
53	I	Iodine	−1, +5		2.13		2.5	^{127}I
54	Xe	Xenon	0					^{128}Xe, ^{129}Xe, ^{130}Xe, ^{131}Xc, ^{132}Xe, ^{134}Xe, ^{136}Xc
55	Cs	Caesium	+1		1.78	1.82	0.7	^{133}Cs
56	Ba	Barium	+2		1.44	1.50	0.9	^{130}Ba, ^{132}Ba, ^{134}Ba, ^{135}Ba, ^{136}Ba, ^{137}Ba, ^{138}Ba
57	La	Lanthanum	+3		1.13	1.26	1	^{138}La, ^{139}La
58	Ce	Cerium	+3, +4		1.09	1.22	1	^{136}Ce, ^{138}Ce, ^{140}Ce, ^{142}Ce
59	Pr	Praseodymium	+3		1.08	1.22	1	^{141}Pr
60	Nd	Neodymium	+3, +4		1.06	1.20	1	^{142}Nd, ^{143}Nd, ^{144}Nd, ^{145}Nd, ^{146}Nd, ^{148}Nd, ^{150}Nd
61	Pm	Promethium	+3		1.04			^{145}Pm
62	Sm	Samarium	+2, +3		1.04	1.17	1	^{144}Sm, ^{147}Sm, ^{148}Sm, ^{149}Sm, ^{140}Sm, ^{152}Sm, ^{154}Sm
63	Eu	Europium	+2, +3		1.03	1.15	1.1	^{151}Eu, ^{153}Eu
64	Gd	Gadolinium	+3		1.02	1.14	1.1	^{152}Gd, ^{154}Gd, ^{155}Gd, ^{156}Gd, ^{157}Gd, ^{158}Gd, ^{160}Gd
65	Tb	Terbium	+3		1.00	1.12	1.1	^{159}Tb
66	Dy	Dysprosium	+3		0.99	1.11	1.1	^{156}Dy, ^{158}Dy, ^{160}Dy, ^{161}Dy, ^{162}Dy, ^{163}Dy, ^{164}Dy
67	Ho	Holmium	+3		0.98	1.10	1.1	^{165}Ho
68	Er	Erbium	+3		0.97	1.08	1.1	^{162}Er, ^{164}Er, ^{166}Er, ^{167}Er, ^{168}Er, ^{170}Er
69	Tm	Thulium	+3		0.96	1.07	1.2	^{169}Tm
70	Yb	Ytterbium	+3		0.95	1.06	1.2	^{168}Yb, ^{170}Yb, ^{171}Yb, ^{172}Yb, ^{173}Yb, ^{174}Yb, ^{176}Yb
71	Lu	Lutetium	+3		0.94	1.05	1.2	^{175}Lu, ^{176}Lu
72	Hf	Hafnium	+4		0.79	0.91	1.3	^{174}Hf, ^{176}Hf, ^{177}Hf, ^{178}Hf, ^{179}Hf, ^{180}Hf

◼ Table 1.1 (continued)

Element			Ionic charge	Ionic radius (Å)			Electronegativity	Major isotopes
Atomic number	Symbol	Name	(Valance state)	cn4	cn6	cn8	Pauling scale	
73	Ta	Tantalum	+5		0.72	0.77	1.5	^{180}Ta, ^{181}Ta
74	W	Tungsten	+6	0.50	0.68		1.5	^{180}W, ^{182}W, ^{183}W, ^{184}W, ^{186}W
75	Re	Rhenium	+4, +6, +7	0.63			1.9	^{185}Re, ^{186}Re, ^{187}Re
76	Os	Osmium	+3, +4		0.71		2.2	^{184}Os, ^{186}Os, ^{187}Os, ^{188}Os, ^{189}Os, ^{190}Os, ^{192}Os
77	Ir	Iridium	+3, +4		0.71		2.2	^{191}Ir, ^{193}Ir
78	Pt	Platinum	+2, +4	0.68	0.71		2.2	^{190}Pt, ^{192}Pt, ^{194}Pt, ^{195}Pt, ^{196}Pt, ^{198}Pt
79	Au	Gold	+1, +3	0.78	0.85		2.4	^{197}Au
80	Hg	Mercury	+1, +2, +3	1.04	1.10	1.22	1.9	^{196}Hg, ^{198}Hg, ^{199}Hg, ^{200}Hg, ^{201}Hg, ^{202}Hg, ^{204}Hg
81	Tl	Thallium	+1, +3		0.97	1.08	1.8	^{203}Tl, ^{205}Tl
82	Pb	Lead	+2, +4	1.02	1.26	1.37	1.9	^{204}Pb, ^{206}Pb, ^{207}Pb, ^{208}Pb
83	Bi	Bismuth	+3, +5		1.10	1.19	1.9	^{209}Bi
84	Po	Polonium	+2, +4		0.67		2.2	^{209}Po
85	At	Astatine	−1, +1		0.62		2.2	^{210}At
86	Rn	Radon	0					^{222}Rn
87	Fr	Francium	+1		1.80		0.7	^{223}Fr
88	Ra	Radium	+2		1.43	1.56	0.9	^{226}Ra
89	Ac	Actinium	+3		1.18		1.1	^{227}Ac
90	Th	Thorium	+4		1.08	1.12	1.3	^{230}Th, ^{232}Th
91	Pa	Proactinium	+4, +5		0.98	1.09	1.4	^{231}Pa
92	U	Uranium	+4, +6		0.97	1.08	1.4	^{234}U, ^{235}U, ^{238}U

The ionic radius is given for the three main coordination numbers (cn) of 4, 6, and 8
Data from Holden [4] and Shannon [7]

varies between 1 and 6 and ionic size between ca. 0.2 Å (e.g., B, C) and ca. 1.8 Å (Cs; ◼ Table 1.1), which means that ionic potential will vary from very low (under 1: very large ions with low charge; e.g., Cs at ~0.6 and Rb at ~0.7) to very high (above 10: very small ions with high charge; e.g., S at ~30 or N at ~40).

1.6.1   Mobility

As noted above, the mobility of a chemical element in the geological environment, and specifically during metamorphism, alteration, and weathering processes—i.e., in presence of free water—is conditioned by its ionic potential. Elements tend to fall in one of the following categories (◼ Fig. 1.10):

— Mobile cations. They don't form very strong bonds with O^{2-} and quickly break from it. As a result, they will move as dissociated ions in aqueous solution, such as K^{1+}, Na^{1+}, Ca^{2+}, Pb^{2+}, and so on.

— Immobile ions. Because of their higher ionic charge, they provide a more focused charge that allows stronger bonds to O^{2-} in solids. Thus they will tend to be immobile during metamorphism and alteration. However, these elements may become somewhat mobile at very high water-to-rock ratios.

— Mobile oxyanions. These elements have the highest ionic charge, which allows them to form very strong bonds with O^{2-} (or other ligands). Accordingly, these elements will be mobile only by associating to oxygen, or other ligands such as BO^{3-}, CO^{2-}, PO^{3-}, SO^{2-}, AsO^{3-}, and so on.

Interestingly, the elements' own characteristics are only one part of the factors controlling their mobility. Other, "external", factors include:

— the availability of oxygen, or the oxidation–reduction potential (Eh) of the system, which will control the valance state of the elements. Let us take Fe as an example: as Fe^{2+} it is mobile in reducing environment

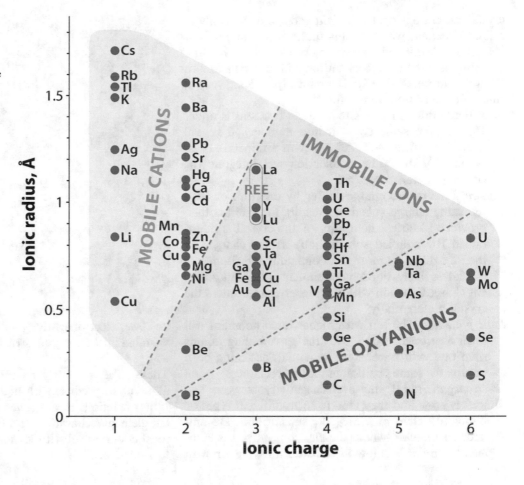

Fig. 1.10 Mobility of chemical elements during metamorphism and alteration as function of their ionic potential. The amount of free water and the prevailing chemical and physical conditions (e.g., oxidation–reduction potential, temperature, available ligands) will also strongly affect the mobility of an element and should always be considered. Based on Goldschmidt [3]

(low Eh), whereas it is immobile, as Fe^{3+} (typically in hematite), in oxidizing environment (high Eh). Uranium is the exact opposite: it is immobile as U^{4+} at low Eh where it forms minerals such as uraninite, UO_2, and is mobile, as U^{6+}, under oxidizing conditions where it moves as various uranyl complexes, such as UO_2^{2+}, $UO_2(HPO_4)_2^{2}$, or $UO_2(CO_3)^{2-}$.

- the availability of other ligands than O, such as Cl, F, or organic acids, in the system;
- the availability of water and the amount of ligands in it;
- temperature: as a general rule, mobility increases with higher temperature.

1.6.2 Compatibility

Another very important consideration is the compatibility of an element with the Fe–Mg silicate crystalline structure. During any igneous process where there is co-existence of a silicate melt and Fe–Mg silicate crystalline structure, an element will preferentially partition into the crystalline phase (compatible) or into the liquid phase (incompatible) (**Fig. 1.11).**

For instance, during fractional crystallization in a magma chamber, the most mafic minerals will crystallize first—predominantly Ca, Fe, and Mg silicates and aluminosilicates—according to Bowen's Reaction Series, and the most compatible elements will tend to partition preferentially into the solid phase, leaving the liquid phase enriched in incompatible minerals. Inversely, during partial melting, the most incompatible elements will tend to leave the silicate crystalline structure and partition preferentially into the silicate melt.

Compatibility is not an absolute quantifiable value, but rather a *relative quality*: at any given point during co-existence of a silicate melt and silicate minerals, the *more compatible* elements will tend to partition into the minerals and the *more incompatible* will tend to partition into the liquid phase. It is an exceedingly useful concept and we will use it extensively later in this book, when we discuss magma differentiation. However, we can already make the important observation that the more mafic and less differentiated a lithology is, more enriched in compatible elements it is, and inversely, more differentiated and more felsic rocks are more enriched in incompatible elements.

1.6.3 Substitutions

The concept of compatibility is, to a certain extent, related to and conditioned by the possibility of element substitutions in a mineral crystal structure. The mineral

1

crystal structure is relatively rigid and can distort only to a limited extent, which means that the size of the ions is a factor. A mineral also has to be electrically neutral, meaning that the charge of an ion will also be a factor. With this in mind, the Goldschmidt's (plus Ringwood's) rules of substitution were defined:

Size: If the difference in size between two ions is under 15% (at the same coordination number), they will readily substitute for each other in a mineral's crystalline structure. The smaller ion will integrate the crystal preferentially.

Charge: Ions whose charges differ by one unit or have the same charge substitute readily for one another, provided electrical neutrality of the crystal is maintained (by coupled substitution). If the charges of the ions differ by more than one unit, substitution is much less likely. For ions of similar radius but different charges, the ion with the higher charge enters the crystal preferentially.

Ionic potential: The ion with higher ionic potential will form a stronger bond with the surrounding anions than one with lower ionic potential, assuming both occupy the same position in the crystal lattice.

Electronegativity: If the electronegativity between two ions is large and they tend to form different types of bonds (covalent or ionic), the substitutions are limited. In the case where the electronegativity is sufficiently similar to allow for substitutions, the ion with

❑ **Table 1.2**   Example of common substitution pairs, based on data from ❑ Table 1.1

Ion 1 (size, in Å)	Ion 2 (size, in Å)	Coordination number
Mg (0.80)	Fe^{2+} (0.86)	6
Al (0.47)	Si (0.34)	4
F (1.25)	O (1.32)	6
Rb (1.68)	K (1.56)	6
Ge (0.48)	Si (0.34)	4
Ga (0.70)	Al (0.61)	6
Hf (0.91)	Zr (0.92)	8
Mn^{4+} (0.62)	Fe^{3+} (0.73)	6
Mn^{2+} (0.91)	Fe^{2+} (0.86)	6
REE (1.26–1.05)	Ca (1.20)	8
Na (1.24)	Ca (1.20)	8

lower electronegativity will be preferentially incorporated into the crystal structure.

These rules are fairly robust and very useful, as they allow us to predict, with high level of certainty, substitution reactions in any given situation, using the different elements' characteristics (❑ Table 1.1). Some of the most common substitutions are given, as an example, in ❑ Table 1.2.

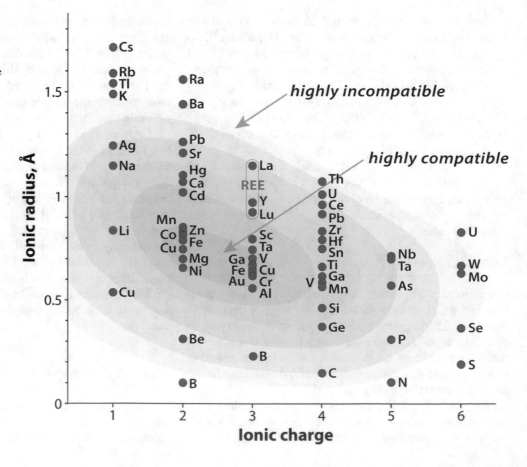

❑ **Fig. 1.11** Compatibility of chemical elements with the crystalline structure of Fe–Mg silicate minerals, relevant during any igneous process where silicate melt and silicate minerals are co-existing

Substitutions are, however, also affected by the physical and chemical conditions of the system. Higher temperature will allow for higher substitution tolerance, because of the thermal expansion of minerals, creating space for larger ions. Inversely, higher pressure will result in lower tolerance for substitutions, as increasing pressure causes compression. Finally, the availability of ions in the system also has a direct effect on substitutions and controls to a significant degree the extent of substitutions that occur during any given geological process.

The effects of pressure and temperature on substitution resulted in the empirical development of two very useful sets of investigation tools: geothermometry and geobarometry. In simple terms, we can estimate the temperature or pressure at which a mineral formed by the amount of substitutions that occurred. One well-known example is called Ti-in-quartz: the higher the amount of Ti-in-quartz, the higher the temperature at which it formed. Another example is the amount of Fe (plus Ga, Ge, Mn, and In) in sphalerite, which can give not only the temperature, but also the pressure of formation of the sphalerite.

1.7 Summary

The chemical elements have their individual characteristics, which strongly affect and condition their behaviour during the various processes in the geological environment. We should always consider these characteristics and behaviours, as well as the different categories to which the elements belong, in particular those specific to geology (reservoir, abundance, and isotopes; mobility in presence of water, compatibility, and substitutions). This chapter also established the vocabulary that we will use throughout the book and can serve as a resource as we discuss specific geochemical cases and applications.

❓ Exercises

Q1.1 Which elements formed in early nucleosynthesis, during the Big Bang?

Q1.2 Choose one of the following elements: Mo, W, Ag, Cd, P, In, Sn, Tl, and Bi. Summarize its characteristics using information from this chapter. Using data in ◘ Table 1.1, comment on the element's geochemical behaviour. What is the likelihood of this element substitution for Fe in olivine and for K in muscovite?

Q1.3 Why are the inert (noble) gases the most stable chemical elements?

Q1.4 Knowing the atomic number of platinum, calculate the number of neutrons in its nucleus, for its different isotopes listed in ◘ Table 1.1.

Q1.5 Using the information in ◘ Table 1.1, find which elements have only one isotope.

Q1.6 Why is the line of stability sloping away from the $N = Z$ line? What happens to a nucleus away from the line of stability?

Q1.7 What is the effect of temperature on the fractionation of stable isotopes?

References

1. Arnett D (1996) Supernovae and nucleosynthesis. Princeton University Press, Princeton, p 11
2. Haynes WM CRC handbook of chemistry and physics: abundance of elements in the earth's crust and in the sea, 97th edition (2016–2017), pp 14–17
3. Goldschmidt VM (1954) Geochemistry. In: Muir A (ed) Clarendon Press, Oxford
4. Holden NE (2004) 11. Table of the isotopes. In Lide DR (ed) CRC handbook of chemistry and physics, 85th edn. CRC Press, Boca Raton
5. O'Neil JR (1986) Theoretical and experimental aspects of isotopic fractionation. In: Valley JW et al (eds) Stable isotopes in high temperature geologic processes, reviews in mineralogy, vol 16. Mineralogical Society of America, Washington, pp 1–40
6. Pauling L (1932) The nature of the chemical bond. IV. The energy of single bonds and the relative electronegativity of atoms. J Am Chem Soc 54:3570–3582
7. Shannon RD (1976) Revised effective ionic radii and systematic studies of interatomic distances in halides and chalcogenides. Acta Crystallographica A 32:751–767

Further Reading

8. **The Elements**, P.A. Cox, Oxford University Press, 1990, ISBN 0-19-855298-X. A very interesting little book, and a perfect first-stop place to learn more about the elements.
9. **Geochemistry**, A.H. Brownlow, Prentice-Hall, 1979, ISBN 0-13-351064-6. It may sound dated, but is a sound and trustworthy source of information. In particular use for us here are Chapters 1 and 5.
10. **Isotopes, Principles and Applications**, G. Faure and T.M. Mensing, Wiley, 2018, ISBN 978-81-2653837–9. This is the definite and complete treatise on isotopes, absolutely complete and up to date. The only drawback is that it's heavy going. It goes into nearly 900 pages and is not really suitable for someone who is trying to get the grasp of the fundamental principles first.

Analytical Methodology and Data Treatment

Contents

2.1   Sampling – 18
2.1.1   Purpose of Sampling – 18
2.1.2   Representativity of Sampling – 18
2.1.3   Sampling Scheme – 18
2.1.4   Quality Control – 18
2.1.5   Size of Sample – 19
2.1.6   Types of Samples – 19
2.1.7   Taking Care of Samples – 20

2.2   Sample Treatment – 21
2.2.1   Sample Preparation – 21
2.2.2   Dissolution – 22

2.3   Most Common Analytical Methods – 22
2.3.1   Atomic Spectroscopy – 22
2.3.2   Quadrupole Mass Spectrometer – 24
2.3.3   Portable XRF – 26

2.4   Measuring Isotopes – 26
2.4.1   Mass Spectrometer – 26

2.5   Microbeam Analyses – 27
2.5.1   Electron Microprobe – 27
2.5.2   Laser Ablation – 28
2.5.3   Secondary Ion Mass Spectrometry – 28

2.6   Data Treatment and Presentation – 28
2.6.1   Purpose of Data Treatment – 28
2.6.2   Single Variable – 29
2.6.3   Anomaly – 30
2.6.4   Two Variables – 31
2.6.5   Multi-variate Analysis – 33

2.7   Summary – 33

References – 34

Electronic supplementary material The online version of this chapter (▶ https://doi.org/10.1007/978-3-030-72453-5_2) contains supplementary material, which is available to authorized users.

© Springer Nature Switzerland AG 2021
P. Alexandre, *Practical Geochemistry*,
Springer Textbooks in Earth Sciences, Geography and Environment,
https://doi.org/10.1007/978-3-030-72453-5_2

2

2.1 Sampling

2.1.1 Purpose of Sampling

The purpose of sampling is, in a nutshell, to collect *representative samples* appropriate for the specific intended purpose of the study. In other words, when planning and conducting sampling we must always keep in mind the purpose and the characteristics of the sampled medium, be it rock or another type of material. There are several very good publications that offer solid and detailed advice on sampling and should be consulted (e.g., Pitard 2019; Esbensen 2020); here we will consider some general ideas and pointers.

2.1.2 Representativity of Sampling

Broadly speaking, there are two types of geochemical studies, those related to scientific research and those related to the exploration and mining industry, even though there is sometimes overlap between the two.

The purpose of the first is to study a particular rock formation, and therefore, we need to collect enough samples to have a clear idea of what the typical representative of this rock is and what is the variability in its characteristics. If, for instance, we study a very homogeneous and uniform granitic intrusion, just a few (maybe fewer than 10) carefully selected samples will be sufficient to represent the whole formation. If, however, lithological variabilities are present (e.g., alterations) and the purpose of the project is to study them separately, then each defined facies needs to be representatively sampled using the appropriate number of samples that captures each facies' own typical examples and their variabilities.

The purpose of exploration, on the other hand, is very different: we are looking for anomalously high values of an element of interest, the commodity we are exploring for. In this case, we need to have a suffi-ciently dense spatial representation to be able to detect an anomalous zone: we are not looking not only for the typical geochemical characteristics of the area, but also for the *significantly different* parts, which will be the targets of further exploration efforts. Necessarily, we need to collect many more samples covering the entire exploration area. (The exact same principle will apply when sampling drill core.)

2.1.3 Sampling Scheme

When planning our sampling campaign, we decide in advance what the strategy will be. With the help of topographic and geological maps and of any available geological information we can gather, we decide on where, how many, and what types of samples we will collect. In the case of sampling for exploration purposes, we will plan the sampling scheme (a regular or irregular grid or a traverse; ◘ Fig. 2.1) and the sampling density. In a typical example, we might decide to cover an area of 20 by 20 km by taking soil samples every 500 m, resulting in 40 lines with 40 samples each, or a total of 1600 samples. If we want to reduce the number of samples to save money, we can decrease the density—and thus reduce resolution and possibly exploration effectiveness—to one sample every 1000 m, resulting in 20 lines of 20 samples each, or 400 samples. In other words, economic considerations are also a constant factor, regardless of the purpose of the study. Simply put, we will always attempt to collect the minimum number of samples that adequately serves our purpose.

2.1.4 Quality Control

Another consideration when sampling for exploration purposes is the reliability of the geochemical data we will obtain. At the sampling stage, we must collect duplicate samples, the purpose of which is to study the reproducibility of the data. A typical scheme includes one

◘ **Fig. 2.1** Example of a sampling scheme typically used in exploration, covering an area of 10 by 10 km, with sampling spacing of 500 m, resulting in 400 samples. On the left, the theoretical planned scheme; on the right, the actual one: some areas might be inaccessible (e.g., lakes, bogs), and the sampling sites might not always be exactly where planned, due to interferences (infrastructure, dense bushes, small boggy areas, etc.)

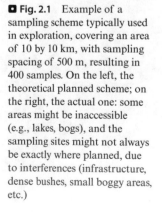
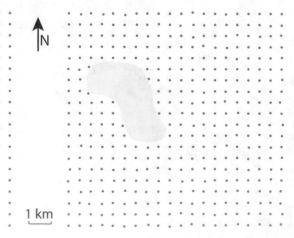

duplicate sample for every 10th sample. Additionally, we are interested in how accurate the results are, as measured against some external sample with a well-known chemical composition, or standard. That standard must be of the exact same type as the material that we are sampling. Typically, approximately 1% of the samples we submit to the laboratory are "external standard". In other words, approximately 11% of our sampling is done uniquely with the purpose of quality control of the data: this might sound wasteful, but is in reality a very necessary procedure to assure that we can trust the data. If all the duplicate samples give the same result (within analytical uncertainty) and if all the standard samples give the known value, then we are that much more certain that the observed values and variations represent the geological reality.

It is worth noting here that these two methods of quality control, measuring external standards and measuring duplicate analysis, are used to examine the *accuracy* of the data (how close they are to *reality*) and the *precision* of the data (how close to *each other* they are; also referred to as *reproducibility*).

2.1.5 Size of Sample

In general terms, the size of the sample will primarily depend on the heterogeneity of the sampled medium, its grain size distribution, and the desired level of precision (Fig. 2.2). In other words, the highest precision will be achieved if we collect a very large sample that is very homogenous and has very low grain size variability. Inversely, the lowest precision—and indeed reliability—will be achieved if the collected sample is very small, very heterogeneous, and has very high grain size variability.

There are several mathematical formulations formalizing these general rules, such as Pierre Gy's fundamental sampling error model, incorporating the material's properties and the desired sampling vari-

ability, and expressed by the following mathematical formulation:

$$m = \frac{Cd^3}{\sigma_s^2}$$

where m is the mass required, C is a number derived from the physical properties of the material, d is the grain size, and σ_S is the sampling precision desired. The C parameter itself is derived from four factors, such as the *shape factor* (from 0.1 for a flat disc to 1 for a cube), the *size distribution factor* (from 0.25 for a wide size distribution to 1 for a narrow size distribution), the *liberation factor* (from 0.1 for unliberated particles to 1 for liberated particles), and the rather convoluted *constitution factor* (depending on the amount and density of the critical particles and the density of the other particles present). In other words, in order to accurately calculate the amount of material required, the physical characteristics of the sampling medium have to be fairly well known, or we will be reduced to guessing. In that case, which is a common situation, the sampler would rely on experience and should err on the side of caution by collecting larger rather than smaller sample, where logistics permit. In principle, the geologist in charge of the sampling programme—whatever its purpose is—should always apply a solid dose of common sense and of careful observation of the sampling media available.

2.1.6 Types of Samples

There are four types of samples that are routinely collected for any purpose: water (stream, lake), rock, soil and till, and vegetation (wood and branches, leaves).

Water is the easiest one to collect and the one that does not require very large amounts, as its heterogeneity is by definition very low. Typically, approximately 100 cc of water is taken from the surface or at depth in body of water in a plastic vial; sometimes additives (e.g., 1%

 Fig. 2.2 Examples of rocks with a different homogeneity and grain size distribution, from a heterogeneous rock with moderate grain size variability (rhyolite, far left) to a very homogeneous rock with very low grain size variability (gabbro; far right). The approximate minimum sample size to obtain a reasonably representative sample, indicated by the dashed line squares, is necessarily different for the different rocks

HNO_3) are put in, to prevent coagulation and separation of solid phases. Additionally, the pH of the water sample can be measured during sample collection (we will see why later, in ▶ Chap. 4).

Rocks are our major object of study and therefore our most common sampling medium. A piece of the rock, larger than the minimum size we have estimated (◘ Fig. 2.2), is broken from an *in situ* rock using a geological hammer or, in some cases, sledge hammer or a rock saw.

Soil sampling is very common in geochemical exploration. From the three main soil horizons, A (top, post-organic matter rich), B (middle, transitional characteristics), and C (bottom, most bedrock fragments rich; ◘ Fig. 2.3), experience has demonstrated that B or C horizons are best suitable for geochemical exploration. Crucially, in order to know at what depth the desired horizon is, the sampling geologists must excavate fairly deeply and expose most of the soil profile. The depth of the soil varies extensively as function of climate (variations in heat and humidity), topography (mountainous vs. plains), and age (juvenile or mature). It is therefore the responsibility of the geologist in charge of the sampling programme to get informed, by collecting the necessary information and conducting a few test holes, about the depth of the horizon that needs to be sampled and to train all geologists in properly identifying it and sampling it.

The soils are fairly heterogeneous and have fairly large grain size variability, which means that soil samples are large, typically between 2 and 5 kg. After the appropriate horizon has been identified, the sample is placed in a plastic bag and sealed. Care is taken to avoid any change in its condition (e.g., drying up), as that may affect the results.

Finally, vegetation is sometimes sampled in geochemical exploration. We can take leaves, branches, trunk wood (◘ Fig. 2.4), or, more rarely, bark. In each case, the sample is placed in a plastic bag, taking care not to affect its integrity.

2.1.7 Taking Care of Samples

It is exceedingly important that we take very good care of the samples collected, as a very large proportion of the observed chemical variability may be due to sampling errors; we want to minimize those as much as possible. Sampling errors fall into different categories, such as
— contamination: the involuntary or voluntary addition of extraneous material into the sample;
— losses through adsorption, condensation, or precipitation;
— changes in the physical and chemical composition and characteristics of the sample.

Any of these errors will change the results, sometimes significantly; therefore, samples must be treated with the

◘ **Fig. 2.3**   Typical soil profile of a moderately mature temperate zone soil. B or C horizons are the most suitable for geochemical exploration

utmost care. As a simple example, companies exploring for gold require that their sampling geologists remove any gold jewellery when conducting soil sampling, in particular rings, as their presence has sometimes been the source of involuntary contamination leading to false anomalies.

There is also the very significant risk of mixing of sample numbers; this is actually the most common sampling error that can lead to significant difficulties in data interpretation. This is why scientists and exploration companies alike have devised methods of keeping track

Fig. 2.4 Sampling of vegetation. Left top, the corer is placed into the tree trunk at approximately chest height; bottom left, the tree core is extracted from the sampling tube (the bark is removed). On the right, spruce crown collected for exploration purposes. Ideally, only one species of tree is sampled in a single campaign

of the samples. For instance, if rocks are sampled, the sample's unique number can be written directly on the rock, copied on the sample bag, and then entered in a master list. Samples can also be grouped in batches—of 20, or 50, or 100—in an attempt to limit any possible mislabelling to a single batch.

2.2 Sample Treatment

In a majority of cases, the geologist will submit the sample to an external laboratory, sometimes sending them directly from the field. That means that all the steps described from here on are conducted by the laboratory personnel. It is however essential that the methodology of sample preparation, treatment, and analyses are well understood, as that will be very helpful with data interpretation. In all cases, it is very useful to discuss about our sampling and analytical programme purpose with representatives from the chosen analytical facility. These people are experienced and helpful and are a very useful resource in selecting the method that is most suitable for any particular purpose; they should be considered a valuable ally.

2.2.1 Sample Preparation

Regardless of the sampling medium and the analytical method, samples are, in the great majority of cases, put

into solution, i.e., dissolved. This solution is then analysed by the chosen method. Before dissolution, a few preliminary preparatory steps are followed, depending on the type of sample.

Rock samples are first crushed, using a jaw or disc crusher, and then sieved to a specific size appropriate for the dissolution method. Importantly, we should be aware that the crushing process can—and sometimes does—introduce contaminants. For instance, crushers equipped with tungsten carbide can lead to significant W or Co contamination; the often-used chrome steel crusher can contaminate the sample with to Cr, etc. Care is taken during sieving not to modify in any way the mineralogical composition of the different size fractions, as that would directly affect the results. At this stage, samples are often "ashed" (heated at high temperature) to remove organic carbon prior to dissolution, which has the benefit facilitating the dissolution process.

Soils are often dried and some external materials, such as pieces of wood, are removed. There is often no need to sieve soil samples, and they can be dissolved such as they are.

When dealing with vegetation, we first select a small subsample to analyse, and then we can follow two possible courses of action: the sample is burnt, and the resulting ashes are dissolved and analysed, or the sample is directly dissolved.

Finally, there is very little sample preparation when the sample is water: the elements of interest are already

2

in solution and can be analysed directly. This is of course not always the case and some preparation could be required, depending on the type of water sample. If, for instance, we are dealing with high-salinity waters (e.g., brines and seawater) we might dilute the sample; or if the elements of interest are present in very low amounts, we might resort to pre-concentration of the water sample, typically through evaporation or ion-exchange chromatography.

2.2.2 Dissolution

Sample dissolution falls into two broad categories: *partial* or *near-complete*, and the choice is dependent on the type of sample and the purpose of the study. To these we might add the less commonly used *sequential leaching*, where we progressively dissolve more and more of the sample, using stronger and stronger reagents, and analyse the solutions separately.

Specifically, the major consideration is whether the element of interest is contained and bound within mineral phases that are constituent parts of the sample collected, or if it is adsorbed and weakly bound to the surface of the sample. In other words, are we interested in the totality of the sample or only in a part of it? If the former, than we need to dissolve as much of the sample as we can, because any part that is not dissolved will not be analysed. Inversely, if we are interested in a component of the sample that is only weakly bound, then we will only partially dissolve the sample in order to preferentially mobilize, or "leach", only the mobile part of the sample; if we dissolve the whole sample we would only dilute the signal from the element of interest.

Partial dissolution, or leaching, is often the method of choice in mineral exploration. The element of interest—the commodity we are exploring for—may have migrated from depth (where the deposit is) to a surface environment (where it forms a surface anomaly). There, the element will often be absorbed on the surface of clay minerals the soil. We are not interested in the chemistry of the clay minerals in themselves, but only in the absorbed phase, in which case we will definitely prefer to employ partial dissolution or leaching. There are several reagents, or leachants, that are routinely used, starting from distilled water, which is perfectly able to mobilize any soluble phases, to NH_4, dilute NH_4OH, HCl, aqua regia (three parts HCl and one part HNO_3), or any combination of these, plus several proprietary reagents.

If, on the other hand, we are interested in the totality of the sample, then it is in our interest to dissolve as much of it as we can. There are two main methods for this. The first is to crush the sample relatively finely and dissolve it with some very strong acids or a combination of strong acids, such the often-used four-acid method

(HF, $HClO_4$, HCl, and HNO_3). It is very important to note, though, that this method will not dissolve the absolute totality of the sample: some resistant ("refractory") minerals will not be fully dissolved. This will have to be taken into account when the results are interpreted.

There is another method, which is more complicated and thus more expensive, but that guarantees that the totality of the sample will be dissolved. It consists of first crushing the sample very finely, to approximately 1 μm, and then heating it to a very high temperature in the presence of a "flux" ($LiBO_4$ or Na_2O_2), which guarantees that all minerals are melted. The sample is then cooled, and the resulting glass is then crushed again and easily dissolved with a strong acid to place all elements in solution.

2.3 Most Common Analytical Methods

There are two fundamental properties of the chemical elements on which the vast majority of analytical techniques are based: (1) the number of electrons and electronic configuration and (2) atomic mass.

The number of electrons of the unbound ion and their configuration are equal to the number of protons in its nucleus. As noted in ▶ Chap. 1, a chemical element is defined by that number, the atomic number, which is directly related to its chemical characteristics and behaviour.

The atomic mass corresponds to the number of protons and neutrons in its nucleus (we will assume that the mass of electrons is negligible). Most elements have more than one isotope with different atomic mass; however, the proportion of the different isotopes for each element is well known.

2.3.1 Atomic Spectroscopy

The atomic spectroscopy family of methods are based on the number of electrons and the unique and specific electronic configuration of the individual elements.

Let us look at an example. Potassium has 19 protons and 19 electrons, placed in 4 different orbitals: 2 at the K shell ($1s^2$), 8 at the L shell ($2s^2 2p^6$), 8 at the K shell ($3s^2 3p^6$), and one at the M shell ($4s^1$); we can write this as $1s^2 2s^2 2p^6 3s^2 3p^6 4s^1$ or simplify it to 2–8–8–1. Crucially, this configuration is unique to potassium: no other element has the same. On the other hand, each of the electron levels have their own and specific energy. If we apply external energy (e.g., heat, light, or radiation) to our sample, part of this energy will be *absorbed* by the sample and used to move electrons from a ground state to an excited state. Importantly, each individual transition of an electron from a ground state to an excited state involves a precise and known

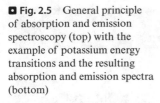

Fig. 2.5 General principle of absorption and emission spectroscopy (top) with the example of potassium energy transitions and the resulting absorption and emission spectra (bottom)

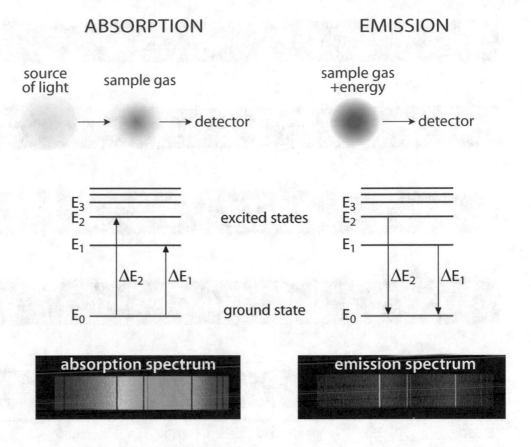

amount of energy being "consumed". In other words, if we observe that a specific amount of energy has been absorbed, we will know that a specific electron has "jumped" from a ground level to an excited level. As potassium has several electrons and each of them can move from a ground level to an excited level, we end up with a specific series of energies that are absorbed by the sample. We will call this an *absorption spectrum* (▣ Fig. 2.5).

Inversely, if we apply external energy directly to our sample, we will take our electrons to an excited level and when they transition back to a ground level, the corresponding transitions ("relaxation") will liberate extra energy, often in the form of visible light (▣ Fig. 2.5); the resulting energy spectrum will be specific and unique to each element (▣ Fig. 2.6).

There are several analytical methods based on this approach, and some of these have been used for a more than a century (▣ Fig. 2.7); they are indeed classical and traditional chemical methods collectively called "spectroscopy". In practical terms, there is a direct relationship between energy and frequency of radiation, more often than not in the visible light spectrum: energy $= h \times$ frequency, where h is Planck's constant $(6.626 \times 10^{-34}$ J s). This is to say that we can easily conceive an analytical apparatus based on this principle, and we will consider some of these here.

An example of a modern development based on this principle is the Optical Emission Spectrometry (OES).

Fundamentally, it is composed of two parts: the plasma torch, where the sample is ionized and heated, and the detector. Let us consider them in turn.

The plasma torch, or inductively coupled plasma (ICP; ▣ Fig. 2.8) is a very important and widely used device, capable of heating the sample (introduced as a very fine mist) to very high temperatures. The plasma is produced by a varying magnetic field that changes its polarity at radiofrequency; the magnetic field is generated by an electric current that goes through a copper coil and changes its polarity at the same frequency (▣ Fig. 2.8). As all the particles are forced by the magnetic field to turn this way and then the opposite way at the very high radiofrequency, they heat up by friction, reaching temperatures of up to 10,000 °C. Because of the way it is generated, the plasma is called inductively coupled plasma (ICP).

The light emitted by the sample when heated to such high temperatures is then analysed by a prism (or sometimes by two prisms at 90° to each other) and detected by a light sensor, very similar to that in a high-resolution digital camera. The specific position of the lines are registered by the electronics and interpreted in terms of concentrations. Because we use plasma torch as our source of energy, the method is called ICP-OES and is fairly common. As noted earlier, there are several other methods based on absorption or emission of energy, collectively known as atomic spectroscopy.

2

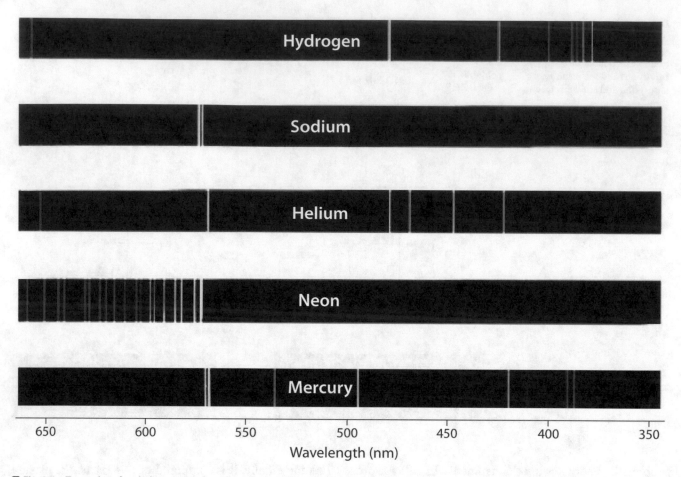

Hydrogen

Sodium

Helium

Neon

Mercury

| 650 | 600 | 550 | 500 | 450 | 400 | 350 |

Wavelength (nm)

▪ Fig. 2.6 Examples of emission spectra for a few selected elements, clearly demonstrating the uniqueness of each spectrum

▪ Fig. 2.7 Famous Tintin—Belgian comic book "reporter" and adventurer—being explained the discovery of a new element by Professor Calys, or Professor Decimus Phostle in the English version (first published in Brussels's *Le Soir* newspaper on 18 November 1941; republished as *L'Étoile mystérieuse* by Casterman in 1942 and as *The Shooting Star* in England by Methuen in 1961). We can see that it is an *emission* spectrum: a meteorite is heated by friction while travelling at high speed in the upper atmosphere and emits visible light, producing a spectrum. The explanation is rather succinct and clear: *"Each of these lines, or each group of lines is characteristic of a metal"*

2.3.2 Quadrupole Mass Spectrometer

Mass spectrometry is another family of methods based on the second characteristic of the elements, their mass. We have two options here, measuring the different isotopes of an element in a sample, which we will consider later in this chapter, or measuring the chemical composition of a sample. For the latter, we will in effect measure a particular isotope and then convert the measurement to the element knowing the proportion of this isotope. For instance, for Fe, we will measure its main isotope, ^{56}Fe, which accounts for 91.75% of Fe, and will then multiply the measured signal by 1.0899 (derived from 100/97.75).

For this method, we use the quadrupole mass spectrometer, coupled with an ICP (▪ Fig. 2.9). This technique is well established, mature, and stable, resulting from five decades of development. It is indeed a very common and widespread analytical method to measure the chemical composition of a geological sample, to the point of being more or less the standard method used by commercial analytical laboratories.

The main principle of quadrupole mass spectrometer is the following (▪ Fig. 2.9). First, the sample is ionized by the ICP and then introduced into

■ **Fig. 2.8** Typical and very common ionization set-up, including sample solution being nebulized (turned into a spray of very fine droplets) and then ionized by the inductively coupled plasma (ICP) "torch". A photograph of actual plasma torch is also shown

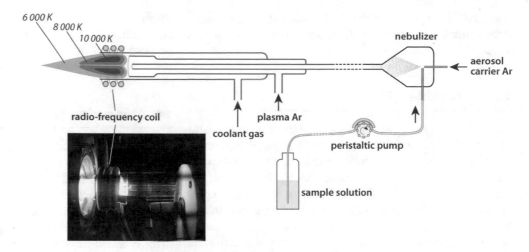

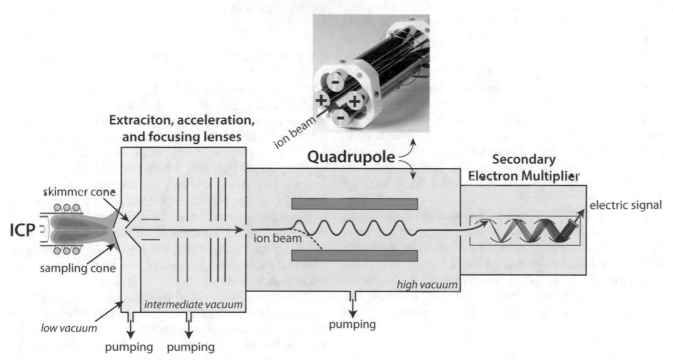

■ **Fig. 2.9** Schematic representation of a quadrupole mass spectrometer instrument, with its main components: The ICP, the electrostatic lens assembly, the quadrupole, and the detector, which is always a secondary electron multiplier (SEM). There are several upgraded and modern variants of this basic design (two quadrupoles, "collision chamber", etc.), but the fundamental principle remains the same

the instrument via small openings in the sampling and skimmer cone. Then the ions are accelerated and focused by a series of electrostatic lenses, resulting in a focused ion beam. The beam then reaches the actual quadrupole, which consists of four small metallic rods, to which direct current is applied, with opposite polarity for the two sets of rods. The polarity is reversed at radiofrequency, making the ions follow an oscillating trajectory. Crucially, we can adjust the voltage in such a way to let only ions with a specific mass travel all the way to the detector (which is always a secondary electron multiplier, or SEM). All ions with different mass will be ejected from the unique trajectory: in effect we filter the ions by mass, which is the definition of mass spectrometer. We can then switch to a different volt-

age, appropriate for a different mass, and measure it, and so on, until all the elements we are interested in are measured.

The quadrupole mass spectrometer (sometimes shortened to just "quadrupole", or even "quad") has several advantages that explain its popularity: it is stable, simple to maintain and use, it is relatively inexpensive, and it is fast: a suite of 50 elements commonly measured in a whole rock analysis can be measured within minutes. This speed is due to the fact that we can change the voltage applied to the rods very quickly and that the SEM is a very fast detector. The instrument has one—relatively benign—shortcoming, which is that it cannot measure isotopic compositions (with the exception of Pb isotopes, and even then not too reliably). For that, we

2

will need another instrument, the magnetic sector mass spectrometer, which we will discuss soon, later in this chapter, in ► Sect. 2.4.

2.3.3 Portable XRF

Let us briefly discuss another analytical instrument capable of measuring the chemical composition of a rock or soil sample, but in situ and in real time: the portable X-ray fluorescence (portable XRF). It relies on the same characteristic of the elements as the atomic spectroscopy: the electrons number and configuration. In this case though, the sample is bombarded by high-energy X-rays or gamma rays, resulting in a brief ionization of the elements in the sample. This makes the electronic structure of the atom unstable, and electrons are transferred from higher to lower orbitals, releasing energy in the form of a photon. Significantly, the energy of this photon is equal to the energy difference of the two orbitals involved, and as we discussed earlier, the unique electronic configuration of each element ensures that the energy of the photon is characteristic of the atoms present in the sample, in a way reminiscent of atomic spectroscopy. There are other methods based on the same principle (e.g., electron microprobe), and we will consider some of them later in this chapter, in ► Sect. 2.5 (◘ Fig. 2.10).

Portable XRF is a relatively recent development of the same technique used in laboratory environment. Being portable is highly valued by exploration geologists as it allows them to obtain chemical information immediately in the field, instead of going through the time-consuming and costly procedure of collecting a sample and sending it to an analytical laboratory. However, we must be aware that the results are never as reliable and repro-

◘ **Fig. 2.10**   Portable X-ray diffraction (P-XRF), shown in use for soil samples analysis

ducible as those obtained by laboratory-based methods such as those discussed above, and must therefore be always considered with a certain share of scepticism (even though the technique is improving quickly).

2.4 Measuring Isotopes

Here we will consider the case when we are uniquely interested in the isotopic composition of a rock or mineral. As mentioned above, we will use the only distinguishing characteristics of isotopes—their mass—given by the sum of the number of protons and the number of neutrons in the nucleus. The specifically designed instrument is called mass spectrometer, as it separates the isotopes uniquely by mass.

2.4.1 Mass Spectrometer

The general principle of the mass spectrometer follows (◘ Fig. 2.11). The first part is the *source*, where the sample atoms are ionized. Very commonly, this is done by the ICP (◘ Fig. 2.8): the first part of the instrument (the very common ICP-MS) will be the same as with the quadrupole (◘ Fig. 2.9), where the ions are introduced into the instrument through the sampling and skimmer cones. In some applications, the sample is gas, and in this case, we use the specifically designed gas source. In both cases, the ions are accelerated and focused into a single ion beam sent flying down the flight tube at high speed (this and subsequent sections of the machine are operating in high vacuum).

Now comes the critical part: the ion beam reaches a strong magnetic field that is generated by a large electromagnet (or a permanent magnet, in some applications). This makes the ion beam deviate from its straight trajectory and curve; the flight tube also curves to accommodate the deviation (◘ Fig. 2.11). Importantly, the heavier isotopes will be that little bit more difficult to deviate from the straight path than the lighter isotopes, due to their higher momentum. After the ion beam passes through the magnetic sector, there will be not just be one ion beam, but several, each made of ions of the same mass. The heavier isotopes will follow a slightly less curved path, and the lighter isotopes will be on a more curved path. In other words, the kinetic energy of different isotopes, directly function of their mass, will allow for the magnetic sector to separate the isotopes of the same element (◘ Fig. 2.11).

We can select which mass will reach the detector by adjusting the strength of the magnetic field by changing the current we use. If we use a permanent magnet, we will vary the speed at which the ions fly by adjusting the acceleration voltage.

Finally, the number of ions is counted in the detector, or collector. Some instruments have more than one

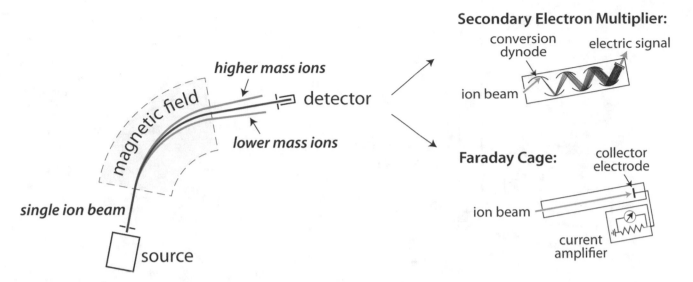

◘ Fig. 2.11 General and simplified schematics of a mass spectrometer, with its three main components: source (where the sample is ionized and accelerated), magnetic sector (where the ions are deviated from the straight flight line according to their mass), and the detector (where the amount of ions are counted; showing the two most common types of detector). There are many variations and developments, but the main principle is always the same: ions are deviated from the straight path according to their mass

detector and can thus measure more than one isotope at the same time; these are called multi-collector ICP-MS (MC-ICP-MS) and are a fairly common instrument, although rather complicated and expensive. They are usually dedicated to purely scientific research and best suited for measuring isotope ratios and not concentrations. There are two major types of detectors: the Faraday collector (also known as Faraday cage) and the secondary electron multiplier (and its derivatives, the channeltron and the microchannel plate, which we will not discuss here).

The Faraday cage is a very simple device: it consists of a small metallic box with an opening on one side and a wire attached to it on the other side. As an ion hits the inside of the box, it will mobilize one or more electrons, thus creating a very weak electric current, which is then amplified (typically 10^{12} or 10^{13} times) and recorded as voltage. The secondary electron multiplier (SEM), is a similarly simple device, consisting of several small curved copper plates arranged in two rows facing each other. An electric potential—typically a few thousand volts—is applied on each plate. The very first plate, called conversion dynode, is the one that the incoming ion beam will hit, generating a small electron beam. That beam will then hit the second plate, situated on the opposite row that generates even more electrons (due to the voltage applied to the plates) and so on. Each subsequent plate will produce even more electrons (that is why we call it multiplier), which will be recorded as voltage. In both cases, the number of incoming ions is proportional to the voltage measured. We have to keep in mind the critical difference between the two types of detectors, namely that the Faraday cage is typically one order of magnitude less sensitive than the SEM.

2.5 Microbeam Analyses

In some cases, it is preferable to analyse only a small portion of a sample, such as when we want to have the chemical—or isotopic—composition of an individual mineral or even different part of a grain. We can, of course, simply extract the grain by crushing the rock and then hand-picking out the specific mineral we are interested in. However, this tends to be labour-intensive and does not always satisfy the requirement of the problematics at hand. In such a situation, we can use a microbeam technique, which generally consists of bombarding the surface of the sample with a focused beam made of electrons or ions or with a laser beam, and then analysing the resulting secondary particles of radiation. There are several microbeam methods based on different types of beams, and we will very briefly discuss three of them here, based on electron beam, laser ablation, and ion beam. This is of course a very succinct and cursory overview of the vast field of microbeam techniques: for more on the topic you might want to visit the excellent *Modern Analytical Geochemistry* (1997) edited by R. Gill.

2.5.1 Electron Microprobe

This is a fairly common instrument, based on the bombardment of the sample surface by a focused beam of electrons (typically between 1 and 5 μm across, achieving very high spatial resolution). As these interact with the chemical elements in the sample, they cause the electrons to be transferred from the ground level to an excited state, as they absorb some of the energy imparted by the electron beam (◘ Fig. 2.12). When the

2

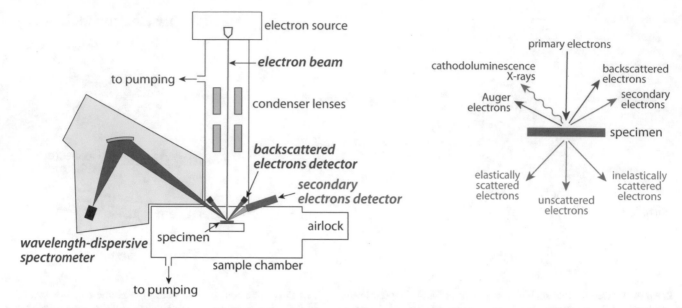

□ **Fig. 2.12** Schematic representation of the electron microprobe (left) and the main interactions occurring when the electron beam impacts the specimen surface (right). Other detectors, such as cathodouminescence detector or an energy-dispersive detector, can also be present in a typical analytical set-up. Commonly, an electron microprobe has four or five wavelength-dispersive detectors, working simultaneously

electrons transition back to a ground level, they emit X-rays with a characteristic energy or wavelength. We will notice that this method relies on the exact same principle as the atomic emission spectroscopy and the X-ray fluorescence: utilizing the specific number of electrons and electronic configuration for each element (□ Fig. 2.5). The method is very reliable and can provide quantitative measurements of any concentrations above a few parts per million (ppm). It is important to note that the "excitation volume" or the part of the sample affected by the energy of the bombarding electrons is significantly larger than the beam itself, typically 10 μm wide and 15 μm deep, making it very hard to measure any grain smaller than 10 μm without the risk of contamination from neighbouring grains.

2.5.2 Laser Ablation

This is a very different technique: the sample surface is bombarded by a laser beam which ablates the sample, basically creating a small crater on it. The laser beam can be anywhere between 10 and 50 μm across, achieving reasonable spatial resolution. The ablated sample is transported, as very small chunks (often smaller than a micron), by a "carrier" gas into the ICP. This means that the same instrument can very often operate in laser ablation *or* solution mode. In the former, the sample is carried into the ICP as a fine dust and in the latter as a fine mist. The laser ablation technique is very often used in conjunction with a quadrupole mass spectrometer for elemental analysis and with a magnetic sector mass spectrometer, for isotopic analysis (LA-ICP-MS).

2.5.3 Secondary Ion Mass Spectrometry

The secondary ion mass spectrometry (SIMS) is a very complicated and therefore expensive analytical technique used only for isotope measurements. It is based on the bombardment of the surface of the sample by a primary ion beam (it can be cations or anions, as suitable for the purpose at hand; in size it is comparable to laser beam, being typically 10–30 μm across). The primary ions create a small amount of plasma and ionise the elements present in the sample. These ions are then extracted, accelerated, and focused by a series of electrostatic lenses to form a secondary ion beam (hence the name) and analysed by an electromagnetic sector following an electrostatic sector, before being measured by the detector, typically a secondary electron multiplier.

This is a complex technique costing very considerable amounts to purchase and operate. In its defence is the fact that it can analyse any isotope of any element at very high spatial resolution in situ, making it invaluable for any sophisticated research task. However, it is hardly necessary for the vast majority of the everyday geochemical applications that a practicing geologist will encounter.

2.6 Data Treatment and Presentation

2.6.1 Purpose of Data Treatment

It is very common in geochemistry to collect a significant amount of data, which can be difficult to communicate and appreciate in their raw form. It is therefore necessary to perform a series of data treatment tasks, the overall purpose of which is to make the data much easier to communicate and understand. If we consider the example shown earlier in this chapter of taking 400 samples on a regular grid (◘ Fig. 2.1), we can see that it would be very difficult to comprehend anything about the data if we consider them without any data treatment. Thus, the purpose of data treatment is to summarize, visualize, and communicate geochemical information.

2.6.2 Single Variable

Although it is exceedingly rare to collect a single variable—the concentration of a single chemical element in the sample—we start by considering the data one sample at a time, before considering them two at a time (correlations) and all together (multi-variate statistics). The first step in our process is to build a histogram, a simple diagram in which the number of samples falling in a particular interval (or "bin") are plotted as function of the concentration (◘ Fig. 2.13).

The histogram is a very powerful tool, as it allows us to comprehend at single glance several critical and significant pieces of information about the data: its overall variation, where most of the data are, and the shape of the distribution. This is why we often start our statistical analysis by producing a histogram for each element analysed.

The second step is to calculate the measures of location for the variable. What we mean by location is where on the *concentration scale* some important markers are situated, specifically those of *central tendency* and of distribution.

There are several measures of central tendency, or where most values tend to be in the geochemical one-dimensional space. The most common of these are the mean, the median, and the mode. The mean is the sum of all values divided by the number of samples. The median is the value found in the middle, when the data are arranged in increasing order; half of the values are above it and half of the values are below it. The mode is the value that occurs the most often (it is on the top of the histogram).

These measures of central tendency have their weaknesses and strengths. The mean is very susceptible to extreme values: just a small handful of extremely low or extremely high values (both situations which are not uncommon in geochemistry) will strongly influence it. Thus, we consider that the mean is not *robust* to extreme values. On the other hand, the mode gives us a very clear representation of where most of the samples are, but does not register the extremely low or extremely high values; it is therefore not a fully faithful representation of the data. It is often considered that the median is the best compromise between these two extremes: it is more robust to extreme values that the mean, and yet takes them into account.

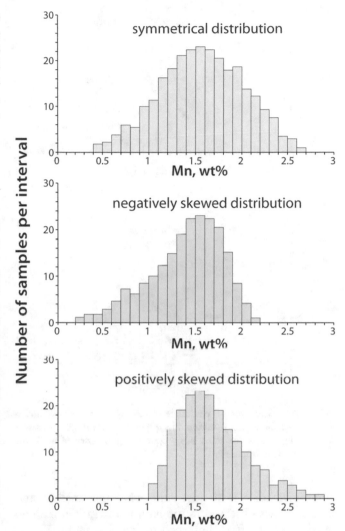

◘ **Fig. 2.13** Examples of experimental histograms, with bin size of 0.1%, showing symmetrical, negatively skewed, and positively skewed distributions. Geochemical data are rarely symmetrically distributed, due to the presence of more than one controlling factor or process

If the distribution of the values is symmetrical, then the three measures of central tendency—mean, median, and mode—will be very nearly the same; but if the distribution is asymmetrical, then we have to make a choice. More often than not the median is the winner (◘ Fig. 2.14). It becomes thus quickly apparent that the choices of the geochemist will affect the outcome of the data treatment. That might sound scary, but it shouldn't be. These choices must always be clearly stated and, if needed, explained and justified. In that way, the audience to which the findings are presented to will always be able to appreciate them for what they are and not be confused. Just saying "On average, there is 1.6% Mn in this rock" is uninformative to the point of being dangerous. We should say "We calculated the median of Mn values, as the distribution is asymmetrical, and it is 1.75%".

We must also make choices—and also clearly state them—on how to represent the data distribution. A very

2

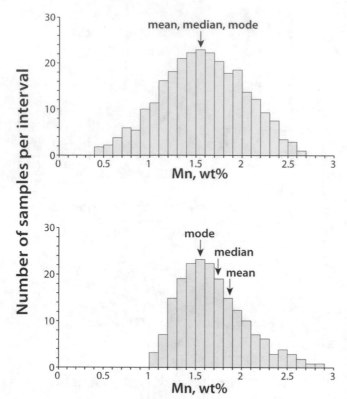

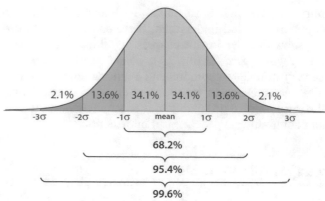

Fig. 2.15 Gaussian distribution and the proportions of the data situated between specific sigma values. We have to exercise caution when we use these proportions, as the geological data are rarely symmetrical or fit a Gaussian distribution (which is, after all, a *probability* distribution)

Fig. 2.14 Measures of central tendency of a symmetrical and a skewed distribution, illustrating the difficulty to faithfully represent with a single value the most likely concentration observed when the distribution is asymmetrical

common choice is to just state what the smallest and largest values observed in the dataset are (minimum and maximum, or often shortened to *min* and *max*). This is a good way to represent variability, but is too sensitive to extreme values (not *robust* enough). Many geochemists prefer to use the 10th and the 90th percentiles: the 10th percentile (P_{10}) is the value below which the lowest 10% of the values collected are; and the 90th percentile (P_{90}) is the value below which the lowest 90% of the values are, when the data are arranged in increasing order. In other words, we report the distribution of only the 80% in the middle of the dataset and "discard" the lowest and the highest 10% of the values. (Again, this is perfectly fine, so long as we clearly inform the audience of what we are doing.). A similar distribution measurement method is the quartiles, where we report the first quartile (Q_1), or the value below which the lowest quarter of the data are, and the third quartile (Q_3), or the value below which the lowest three quarters of the data are, when the data are arranged in increasing order. We can also just report the interquartile range (Q_1–Q_3), or the two values between which the middle 50% of the observations are.

There is another measure of dispersion, the standard deviation, often denoted by σ (sigma), calculated as the "mean deviation from the mean", which is very useful and often used. Its usefulness comes from the possibil-

ity to estimate how much of the data is situated between specific sigma values (■ Fig. 2.15). Even though geochemical values are rarely perfectly symmetrical or can be approximated by a Gaussian distribution, we can estimate that approximately two-third of the data are situated between -1σ and 1σ (or $\pm 1\sigma$), and that virtually all data are situated between -3σ and 3σ (or $\pm 3\sigma$; ■ Fig. 2.15). In that manner, when we report the data distribution in terms of standard deviation, the reader will know approximately how much of the data are comprised within this interval.

Once the measures of central location and of distribution have been calculated, they can be communicated as numbers, with the effort to be succinct and clear. For example "The median value of FeO_2 in these samples is 2.45% and the Q_1–Q_3 range is from 1.97 to 3.23%", or "The mean SiO_2 value for this population is $62.3 \pm 2.5\%$ (1σ)". Here, we have clearly indicated what location measures we have used (median and quartiles in the first example, mean and standard deviation in the second), and then given the values. We will also very often give the results in graphic manner, as a small image (■ Fig. 2.16), which allows us to present several variables together and compare them visually.

2.6.3 Anomaly

A very common and often perplexing topic is defining what is different, or anomalous, or outlying. We can probably all agree on a general and clear definition that an *anomaly is a value or a population of values that is significantly different from the local background or from the majority of the values*. The critical point here is "significantly different", and there is absolutely no agreement in the geological community on what constitutes significantly different, how to define it and how to measure it. People and organizations have their own preferences

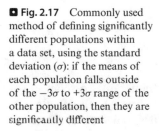

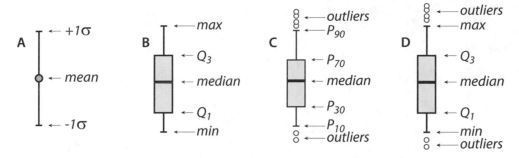

▫ Fig. 2.16 A Few possible visual representations of central tendency and dispersion of a single variable. In A, a simple representation which assumes symmetrical distribution; in B, a very commonly used representation—called "box-and-whiskers"—providing more specific information; in C, another possibility using percentiles and including "outliers". The plot in D represents an internal contradiction and should *never* be used: *min* and *max* are defined in an absolute manner, and there cannot be any values beyond them. Any of the first three visual representations are acceptable; regardless of which one we prefer, we must always clearly indicate which location measures we have used

▫ Fig. 2.17 Commonly used method of defining significantly different populations within a data set, using the standard deviation (σ): if the means of each population falls outside of the -3σ to $+3\sigma$ range of the other population, then they are significantly different

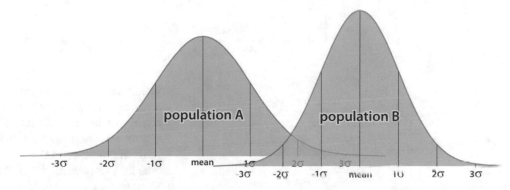

and procedures, and sometimes defend them with unreasonable zeal. Ultimately, it doesn't really matter how we define what is different: whichever method we use we have to explain it and clearly describe it.

We can use relatively simple measures of anomalies or outliers, such as percentile values. We can say, for instance, that any value below the 10th percentile or above the 90th percentile is an outlier (▫ Fig. 2.17). We can also use the standard deviation (σ) value and define an outlier as, for instance, any value lower than mean-2σ or higher than mean $+2\sigma$. Significantly, as anytime we treat our data in any way, we must clearly indicate and define our measures so that the reader can understand the meaning of the results.

As mentioned above, there is no universally accepted statistical definition of anomaly. However, a significant proportion of geoscientists use the standard deviation to define anomaly as something beyond the -3σ or $+3\sigma$ values, meaning that the lowest 0.2% and the highest 0.2% of the data can be considered anomalous. In a similar manner, two data populations may be considered significantly different if their means fall outside of each other's -3σ to $+3\sigma$ ranges (▫ Fig. 2.16).

2.6.4 Two Variables

When dealing with two variables, the mostly important consideration is always how related to each other they

are, or how they correlate. Basically, they can positively correlate (when the values of one variable increase, so do the values of the other variable), negatively correlate (when the values of one variable increase, the values of the other variable decrease), or they may be independent, or uncorrelated (the variation of values for one variable are not related to the variation of values for the other variable). We can also place modifiers (e.g., weakly correlated, strongly correlated). The best visual way to appreciate the interdependence of two variables is to plot them against each other in a simple binary plot (▫ Fig. 2.18).

Regardless of the purpose of the geochemical work done, we are very often very interested to know if two variables are interdependent, which can be visually appreciated by plotting them against each other (▫ Fig. 2.18). However, we are also interested in the degree of interdependence of that relationship. In other words, we want to have a numerical value telling us how strong the relationship between the two variables is; we will call this number the correlation coefficient. It is relatively easily calculated as the average of the vertical deviations (for the variable Y) from the straight line that best fits the data (a method developed by mathematician Carl Friedrich Gauss). The resulting number, often called r, can vary between −1 (the strongest negative correlation) to 1 (the strongest positive correlation), with 0 meaning absolutely no correlation between the two variables.

▣ Fig. 2.18 Graphic representation of the most common cases of the interdependence of two variables

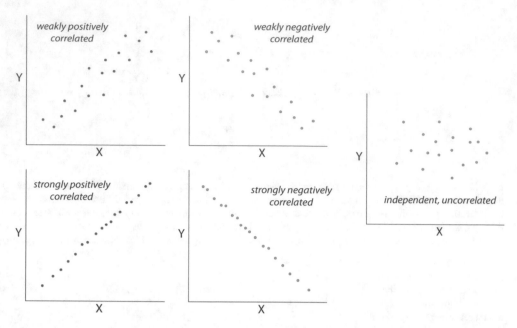

▣ Fig. 2.19 Example of a strong nonlinear correlation between two variables. The Pearson correlation coefficient will underestimate the degree of interdependence between the two variables

It will have not escaped your attention that the correlation coefficient that we just described (also known as the Pearson product-moment correlation coefficient, or simply Pearson) is best suited for linear correlations, when the relationship between the two variables is best described by a straight line (▣ Fig. 2.18). However, this is not always the case: sometimes the relationship between the two variables is not linear but is best approximated by a curve (e.g., polynomial, quadratic, exponential; ▣ Fig. 2.19). In such cases, the Pearson correlation will actually be misleading: even if there is a strong correlation, the calculated correlation coefficient will be low and give the impression of low degree of interdependence between the two variables.

In such a case, we have two options. Firstly, we can calculate a correlation coefficient in a manner similar to that of linear correlation, but using a curve to approximate the relationship between the two variables, which is possible even though cumbersome. We don't always have the luxury of time to define the shape of correlation for each two variables combination, particularly if we deal—as we often do—with a data set composed of many variables. Secondly, we can use another cor-

relation coefficient, the Spearman rank correlation (or simply Spearman, often denoted as r_s). To calculate it, we will not use the actual values, but their rank within the data set when they are arranged in increasing order. In this manner, we make abstraction of the values themselves and only consider if the relationship between the two variables is monotonic. This measure of correlation is more *robust* than the Pearson linear correlation coefficient. In other words, if we know that the correlation is linear we can safely use the Pearson correlation coefficient; but if not, then we are better off using the Spearman rank correlation. This should remind you of something: just as the median is more robust measure of central tendency than the mean (it is less sensitive to extreme values), so the Spearman is more robust than the Pearson correlation coefficient (it is less sensitive to different shapes of interdependence of two variables.)

We mentioned that we are interested in the extent, or degree, of correlation, or interdependence between two values. We also want to know what values of the correlation coefficient, be it r or r_s, are significant. In other words, what is the correlation coefficient value that means that there is real interdependence between the two values? As an example, we might *arbitrarily* decide that an r from 0.8 to 1 means strong correlation, from 0.5 to 0.8 weak correlation, and below 0.5 no correlation (and the same numbers, but with the minus sign, for inverse correlations). However, this is not a good way to approach the problem, as it disregards anything about the samples or the data set. In order to calculate the significant correlation coefficient—and there are several good methods to do that—we should consider two factors: the number of analyses (the higher the number, the lower the significant correlation coefficient will be) and the level of confidence we decide on (a higher confidence level will require a higher significant correlation coeffi-

◘ Table 2.1 Example of correlation table for electron microprobe analyses of uraninite samples, with both Spearman rank correlation (in bold) and Pearson product-moment correlation (in italics)

	UO_2	PbO	SiO_2	FeO	CaO	TiO_2
UO_2	1					
PbO	**−0.89**/*−0.82*	1				
SiO_2	**0.39**/*0.50*	**−0.52**/*−0.62*	1			
FeO	**0.51**/*0.46*	**−0.67**/*0.62*	**0.46**/*0.32*	1		
CaO	**0.58**/*0.65*	**−0.72**/*0.82*	**0.44**/*0.63*	**0.70**/*0.62*	1	
TiO_2	**0.00**/*−0.07*	**0.05**/*0.08*	**0.45**/*0.44*	**−0.18**/*−0.21*	**−0.24**/*−0.14*	1

Data from [1]
There are a few significant differences between the two correlation coefficients, such as SiO_2–CaO and SiO_2–UO_2, showing that the correlations are not always linear

cient). For example, the significant correlation coefficient for the same confidence level of 90% will be about 0.57 for 20 samples and about 0.48 for 30 samples; if we increase the confidence lever to 95% then the significant correlation coefficient will be 0.71 for 20 samples and 0.63 for 30 samples.

2.6.5 Multi-variate Analysis

In the very common case when several—sometimes more than 50—variables are present in the data set, we need to have a straightforward and clear way to summarize the relationships between them. A traditional way is to simply provide the correlation table (◘ Table 2.1). However, when many variables are present, the table can be very large and it becomes difficult to appreciate the *groups of elements* that tend to correlate with each other, which is very often related to and conditioned by some critical processes affecting the rocks samples analysed. There are several statistical methods specifically devised to interrogate the data set and discover the groups of elements behaving in a similar manner, and we cannot consider all of them here. We will just give one example, the very commonly used principal component analysis (PCA), sometimes synonymous with the factorial analysis (FA). This method is based on the individual correlations existing between all of the variables—chemical elements in our case—such as can be found in a correlation table and, through a series of mathematical transformations, creates a graphic representation of the elements' groups and mutual relationships (◘ Fig. 2.20). In general terms, the closer two or more variables are to each other in the PCA graphic results (e.g., UO_2, CaO, and FeO in this example), the more strongly correlated they are; two or more variables at opposite sides (e.g., UO_2, CaO, and FeO on one side and PbO on the other), the more anti-correlated they are; and when two variables or groups of variables describe a right angle between each other when going through the origin, they are independent.

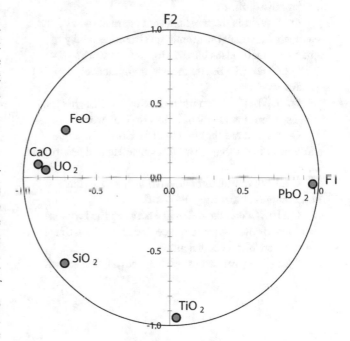

◘ Fig. 2.20 Example of the results of principal component analysis (PCA) for the same data presented in ◘ Table 2.1 [1]. We can notice right away that UO_2, CaO, and FeO belong to the same group and behave in the same way in thus data set; this group is anti-correlated to PbO, while TiO_2 is independent from them

2.7 Summary

When conducting sampling work in the field, we have to have a clear purpose in mind, which will condition the sampling medium (rock, soil, vegetation, water) and the sampling scheme. We must take the utmost care of the samples, as the bulk of errors and confusion happen at the sampling and sample preparation stage. The samples are prepared (crushed, sieved, and dissolved), and analysed, using their major characteristics such as their number of electrons and electron configuration (atomic spectroscopy family of methods; portable X-ray fluorescence; electron microprobe), or the mass of their isotopes (mass spectrometry family of methods).

Once the data are obtained, we have to convey their meaning in a clear and straightforward way. We can do so by providing the central tendency (mean, median, or mode) and dispersion, for either single variable, correlation, two variables, groups of elements, or for multi-variable. In each case, we must clearly state and justify—if needed—the method used and present the findings both as numerical values and as graphic representation.

❓ Exercises

Q2.1 What is the most important purpose of sampling?

Q2.2 What are the major sampling media that we use in geochemistry?

Q2.3 What is the sample size dependent on?

Q2.4 Why do we dissolve the samples and how do we achieve dissolution?

Q2.5 What is the most relevant characteristic of the elements on which the atomic spectroscopy family of analytical methods is based? How does it work?

Q2.6 Describe briefly the way an inductively coupled plasma works.

Q2.7 What is the characteristic of the elements that mass spectrometers rely on? How do they work?

Q2.8 What are the two characteristics of a single variable data set that we need to calculate and report in all cases?

Q2.9 What is an anomaly and how can we know if something is statistically different?

Q2.10 What is the difference between the Pearson Product-Moment correlation coefficient and the Spearman Rank Correlation coefficient?

Q2.11 How can we know that a correlation is real?

References

1. Alexandre P, Kyser K (2005) Effects of cationic substitutions and alteration in uraninite, and implications for the dating of uranium deposits. Can Mineral 43:1005–1017

Further Reading

2. There are several very good books on the topics of sampling, sample preparation, and analysis; some of which are listed below.

3. **Modern Analytical Geochemistry**, R. Gill, Editor, Longman, 1997, ISBN 0-582-09944-7. An excellent book, to be used extensively and without moderation.

4. **Isotopic Analysis: Fundamentals and Applications Using ICP-MS**, F. Vanhaecke and P. Degryse, Editors, Willey, 2012, ISBN 978-3527328963.

5. **Stable Isotope Geochemistry**, J. Hoefs, Springer-Verlag, 1997, ISBN 3-540-61126-6. Chapter One contains some important pointers.

6. **Using Geochemical Data**, H. Rollinson, Pearson Education, 1993, ISBN 0-582-06701-4. A very solid book, with plenty of relevant information. Here we will be particularly interested in Chapters One and Two.

7. **Applied Geostatistics**, E.H. Isaaks and R.M. Srivastva, Oxford University Press, New York, ISBN 019-505012-6. This is a very lucid and useful book. Chapters One and Two are the most relevant here.

8. **Theory of sampling and sampling practice**, F. Pitard, Chapman and Hall/CRC Boca Raton, FL (2019), ISBN 978-1-351-10592-7.

9. **Introduction to the Theory and Practice of Sampling**, K. Esbensen, IM Publications, Chichester, UK (2020), ISBN 978-1-906715-29-8.

Lithogeochemistry

Contents

3.1   Introduction: Definition and Applications – 36

3.2   Whole Rock Classifications and Variability – 36
3.2.1   Classification Diagrams – 36
3.2.2   Classification Diagrams for Igneous Rocks – 38
3.2.3   Classification Diagrams for Sedimentary Rocks – 38

3.3   Variation Diagrams – 39

3.4   Discrimination Diagrams: Tectonic Setting and Mantle Reservoirs – 40
3.4.1   Tectonic Provenance – 40
3.4.2   Mantle Heterogeneity – 41
3.4.3   Magma Differentiation, Source of Sediments, and Degree of Alteration – 45
3.4.4   Magma Differentiation Through Partial Melting and Fractional Crystallization – 45
3.4.5   Rare Earth Elements as Proxy for Magma Differentiation – 47
3.4.6   Rare Earth Elements in Sedimentary Rocks – 48
3.4.7   Quantification of Hydrothermal Alteration – 50
3.4.8   Pearce Element Ratios – 51
3.4.9   Mixing Calculations – 53

3.5   Geochronology – 54
3.5.1   General Principles of Geochronology – 54
3.5.2   Radiogenic Product Retention and Closure Temperature – 55
3.5.3   Correction for Common Lead – 56
3.5.4   Concordia Diagram – 57

3.6   Summary – 58

References – 59

Electronic supplementary material The online version of this chapter (▶ https://doi.org/10.1007/978-3-030-72453-5_3) contains supplementary material, which is available to authorized users.

© Springer Nature Switzerland AG 2021
P. Alexandre, *Practical Geochemistry*,
Springer Textbooks in Earth Sciences, Geography and Environment,
https://doi.org/10.1007/978-3-030-72453-5_3

3.1 Introduction: Definition and Applications

Lithogeochemistry is, in general terms, the study of the *complete* chemical composition of *whole* rocks, considering the main and trace elements, but also isotopes. This is why it is sometimes called *whole rock geochemistry*. Ideally, we will use analytical methods that will provide us with the complete or near-complete whole rock compositions, such as four-acid near dissolution or LiB_5 fusion, as discussed in the previous chapter. If this condition is not satisfied and the data don't faithfully translate the chemical composition of the totality of the rock—such as those obtained by partial dissolution analytical methods—we can still use the data and apply the methods that we will consider in this chapter. However the interpretations must be subject to a reasonable amount of scepticism, reflecting how close the data represent the full, or complete, chemical composition of the rocks.

There are three main fields of application of lithogeochemistry, and we will discuss these in turn:

— Classification of igneous rocks—both intrusive and extrusive—based on their chemical composition;
— Determination of the geotectonic context (or setting) of rocks, and the main Earth reservoirs where they formed;
— Determination and characterization of the processes that affected rocks during (e.g., magma differentiation) or after (e.g., alteration) their initial formation.

These three applications of geochemistry are interrelated: they all derive and relate to the various processes during the evolution and differentiation of igneous rocks, and to the geochemical behaviour of the elements during these processes.

To these three applications we may add the use of lithogeochemistry in mineral exploration and geochronology. Mineral exploration is a large topic to which the next chapter will be dedicated, whereas geochronology is the art of determining when a particular event affecting a rock occurred. We will conclude this chapter with this topic.

3.2 Whole Rock Classifications and Variability

This is the first and most common application of lithogeochemistry. Every time we obtain whole rock geochemical data, the first thing we will do is use them to ascertain the rock type of the samples analysed, based only on the major elements' concentrations. It may be argued that we can discover the rock type by visual observation; this is correct, to a point. Very many factors might conspire to prevent the correct visual determination of rock type, such as alterations and weathering,

metamorphism, outcrop visibility, or even human error. Often, two rocks may look very much the same but be chemically different, in which case geochemistry will be able to tell them apart. Using geochemistry will also provide us the ability to discover subpopulations within the same rock (which may not be possible by observation alone) and appreciate the variability within a rock unit.

3.2.1 Classification Diagrams

The major and most used diagram for igneous rocks is the $(K_2O + Na_2O)$ versus SiO_2, also called total alkali versus silica, or TAS, diagram ([7, 22, 44], ◘ Fig. 3.1). It is the most useful and used classification diagram for igneous rocks, as it is the only one to provide a simple and straightforward igneous rock classification. In any given situation when dealing with the whole rock geochemistry of igneous rocks, this is definitely the first diagram we should turn to. Sometimes the SiO_2 versus K_2O diagram is also employed, but much more rarely as its usefulness is more limited.

To start with, the Si content of the rocks will divide them into ultramafic ($SiO_2 < 45$ wt%), mafic ($45 < SiO_2 < 52$ wt%), intermediate ($52 < SiO_2 < 63$ wt%), and felsic ($SiO_2 > 63$ wt%; Pecerillo and Taylor [26]; ◘ Fig. 3.1). The fields corresponding to different rocks types defined in the diagram can apply to both volcanic and plutonic igneous rocks, even though the fields may differ a bit. (It is slightly less useful for volcanic rocks, due to the slightly higher mobilization of alkalis during weathering and metamorphism.) We can thus, at a glance, have a clear idea of the rock type of our samples and of the geochemical variation present. Finally, we can distinguish between alkaline and subalkaline (or tholeiitic) series of igneous rocks, the latter being by far the more common on Earth (◘ Fig. 3.1).

Other diagrams are also sometimes used for whole rock classification of igneous rocks. A relatively common one is based on the normative mineralogy of the rocks, calculated from the whole rock concentrations and plotted in molecular proportion diagrams (e.g., the albite–orthoclase–anorthite ternary diagram; [3]). Another approach is to use the cation compositions of the rocks, with fields clearly defined in a manner similar to the TAS diagram (e.g., the $R1 = 4Si-11(Na + K)-3(Fe + Ti)$ vs. $R2 = Al + 2 Mg + 6Ca$ diagram; [9]). Both approaches have their usefulness and shortcomings, and may contribute to the understanding and classification of the igneous rocks at hand. However, the TAS diagram remains the most reliable, straightforward, and useful, and is thus the most widely utilized.

Let us consider an example of the use of the TAS classification diagram. A variety of Precambrian intrusive and sedimentary rocks has been described near the town of Bissett (Manitoba, Canada). These range from mafic to felsic in composition and have been described

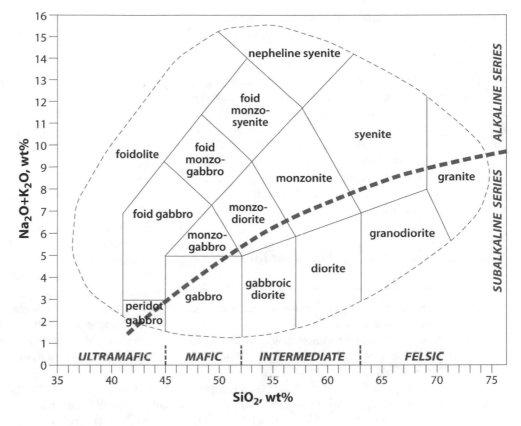

Fig. 3.1 Total alkali (Na₂O+K₂O) versus silica (SiO₂), or TAS, diagram, showing the fields (after [44]) of all major igneous rock types (plutonic in this case), the four main degrees of differentiation (ultramafic, mafic, intermediate, and felsic), and the approximate empirically derived line separating the predominant subalkaline series from the alkaline series. This is the single most useful rock classification diagram based on whole rock major element compositions

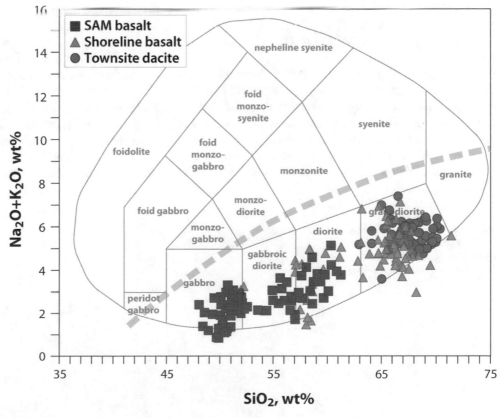

Fig. 3.2 TAS classification of igneous rocks from the Bissett area (Canada), showing the variability of lithological types and revealing the presence of subcategories, difficult to detect by field observations only

and classified based uniquely on field observations. A recent large-scale geochemical effort provided their chemical compositions (☐ Fig. 3.2) and helped redefine their classification.

Firstly, we can see that all the igneous rocks in the area are subalkaline and define a single trend, suggesting a common source and evolution. Secondly, the two basalt units have in fact much more variable compo-

3

sition: the SAM basalt has two subpopulations, one purely basaltic in composition and the other intermediate. The shoreline basalt is not basalt at all, but has two groups with intermediate and even somewhat felsic composition. Only the townsite dacite has a chemical composition corresponding to the visually defined rock type. Once these findings were known, it was possible to refine the lithological description: subtle mineralogical and textural differences were observed between the two SAM basalt groups, for instance. This is a typical example of how the chemical composition of the rocks in an area can significantly improve their classification, when visual observations were difficult and thus somewhat incorrect.

3.2.2 Classification Diagrams for Igneous Rocks

We can further refine the classification of igneous rock using a variety of diagrams designed for a specific rock type. As an example, the Al saturation diagram (◘ Fig. 3.3) defines the main subtypes of granite chemical composition.

On the basis of Al, alkalis, and Ca, it is possible to define the following granitic compositional subgroups:
- Peraluminous: $Al_2O_3 > (Na_2O + K_2O + CaO)$;
- Metaluminous: $Al_2O_3 < (Na_2O + K_2O + CaO)$ but $Al_2O_3 > (Na_2O + K_2O)$;
- Subaluminous: $Al_2O_3 = (Na_2O + K_2O)$;
- Peralkaline: $Al_2O_3 < (Na_2O + K_2O)$.

Another useful diagram is the AFM diagram, or the alkalis–iron–magnesium ternary diagram (◘ Fig. 3.4), on which the two main igneous rock series can be placed, with specific locations for the various degrees of differentiation for individual rock types.

The AFM diagram chemically defines two igneous rock series (subalkaline and calc-alkaline; ◘ Fig. 3.4).

We can consider the TAS classification diagram and its two main series (alkaline and subalkaline) but also the Al saturation diagram (used mostly for granitoids; ◘ Fig. 3.3), and we can define three major igneous rock groups. Crucially, these are likely to be found in different tectonic settings and are formed by different processes:
- The most common **subalkaline group**, with rocks which tend to be Fe-rich, alkali-poor, and peraluminous. The rocks from this group are commonly found in mid-oceanic ridges, intraplate volcanoes, and sometimes convergent margins, and typically correspond to de-compressional melting.
- The **calc-alkaline group**, with rocks that are more alkaline, Ca-rich, and magnesian, are often metaluminous to peraluminous, and are mostly found in convergent margins (subduction). These rocks tend to correspond to partial melting due to the presence of volatiles.
- The **alkaline group**, with Fe-rich and alkali-rich rocks that are often metaluminous to peralkaline. These rocks are found in intraplate situations or convergent margins (continental collision). They frequently correspond to partial melting due to increase of temperature.

In this way, we can not only reliably classify the different igneous rock types, but also define, with a fair degree of certainly, their tectonic context. We will discuss this in a little bit more detail later in this chapter.

3.2.3 Classification Diagrams for Sedimentary Rocks

Metamorphic rocks are not often classified geochemically, but rather as function of their degree of metamorphism and the corresponding mineralogical modifications. Their chemical composition will often closely resemble that of the protolith, with the exception

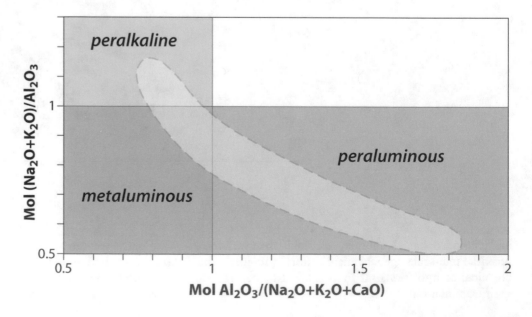

◘ **Fig. 3.3** AL saturation diagram for felsic rocks, defining the three possible compositional types. The empirically derived field of most likely compositions is outlined in dashed lines

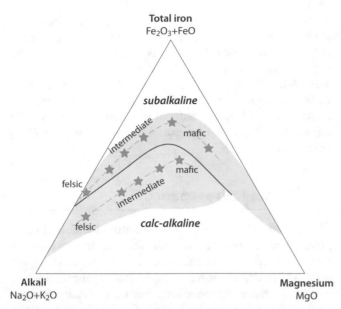

◘ Fig. 3.4 AFM ternary diagram, with the typical compositional fields for two main igneous rock series, defined here as subalkaline and calc-alkaline, their main differentiation trends (dashed lines), and location of rocks with different degrees of differentiation (stars)

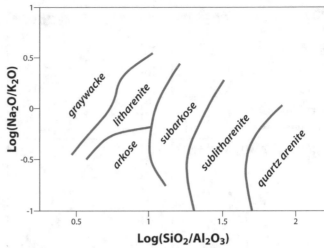

◘ Fig. 3.5 Classification of detrital sandstones, showing the approximate fields of rocks with increasing maturity reflected by higher SiO_2/Al_2O_3 ratios (after [27], and [18], modified)

of increasing devolatilization with increasing metamorphic grade. Orthometamorphic rocks (those derived from an igneous protolith) with low degree of metamorphism will have chemical compositions very similar to those of the protolith, and thus the same classification diagrams can be used (such as the TAS diagram).

Something similar is also true for sedimentary rocks, which are not often classified based on their chemical composition, but rather on their mineralogy, grain size, and texture. However, whole rock geochemistry has its uses here, not only for classification purposes, but also for determining their degree of maturity [33]. It is reflected in the SiO_2/Al_2O_3 ratio, which mirrors the relative amounts of quartz (Si) and of clays and feldspars (Al). This ratio will increase with a greater degree of sandstone maturity and higher degree of destruction of feldspars and clay minerals. This can be extended by adding the Na_2O/K_2O ratio ([27], ◘ Fig. 3.4), which is very widely used. Other sandstone classification diagrams are also used, such as Na_2O/K_2O versus SiO_2 [24] or $log(Fe_2O_3/K_2O)$ versus $log(SiO_2/Al_2O3)$ [18], which have their usefulness.

3.3 Variation Diagrams

As mentioned earlier, we are interested in not only classifying rock (mostly igneous) using whole rock geochemistry, but also examining their chemical variability. Importantly, we want to detect and study the possible relationships between individual rocks within the same area of province, as well as the relative degree of differentiation (for igneous rocks) or maturity (for

sedimentary rocks) of the individual rock units. For this purpose we very often use the so-called variation diagrams, where the chemical elements are plotted against a common variable such as SiO_2 (the clearest indicator of magma differentiation or of maturation of sandstones). These are collectively called Harker diagrams after the geologist who first used them more than a century ago, and we will see a few examples here.

The first example comes from Australia, where the major element compositions of several sandstones of varying maturity were plotted versus SiO_2 (◘ Fig. 3.5). The interpretation here is that increasing SiO_2 reflects the higher proportion of quartz and the decreasing of other major elements reflects the progressive destruction of feldspars, lithic fragments, and other detrital minerals [4] (◘ Fig. 3.6).

The second example shows a suite of igneous rocks with the major elements plotted against SiO_2 (◘ Fig. 3.7). In this case, the chemical evolution observed reflects the continuous magmatic differentiation and evolution of these rocks, from mafic ($SiO_2 < 52$ wt%) to felsic ($SiO_2 > 63$ wt%), likely mostly through a series of fractional crystallization episodes. Considering uniquely their chemical composition, we can infer that the individual rock samples very likely belong to the same major magmatic series (subalkaline in this case).

As a reference, the major and trace element compositions of some major Earth reservoirs (Bulk Silicate Earth, or BSE, upper mantle, and continental crust) and rocks representing the four major degrees of differentiation (ultramafic, mafic, intermediate, and felsic), in addition to the C1 chondrite meteorite, are provided in ◘ Table 3.1. Clear variations can be observed in both the major and trace elements when the differentiation (from ultramafic through to felsic) is considered, reflecting the trends described graphically in ◘ Fig. 3.7.

3

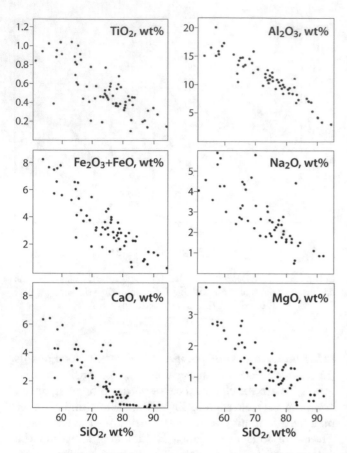

● **Fig. 3.6** Variation (Harker) diagram for a series of sandstones from eastern Australia (after [4], modified), showing the increase of SiO₂ and decrease of all other major elements with increasing maturity

However, we should exercise a fair bit of caution when using Harker diagrams, in particular when SiO_2 is used on the common X-axis. The problem is that clear trends visible on a variation diagram may not correspond to real geochemical relationships, but be artificially produced, due to the unit sum constraint, or the expression of concentrations (in both minerals and rocks) as percentages combining to 100% (Rollinson 1993). This is the reason why Harker, or variation, diagrams are discredited in the eyes on some geochemists while many others continue to use them extensively, albeit with caution. The best way to deal with this problem is to consider the trends observed in a Harker diagram uniquely in the light of a specific hypothesis. In other words, if the observed trends and variations correspond to a theoretical model, then the hypothesis tested—corresponding to that model—is likely to be true (Rollinson 1993).

3.4 Discrimination Diagrams: Tectonic Setting and Mantle Reservoirs

As we mentioned earlier, the chemical composition of an igneous rock will be, to a degree, a function of the tectonic environment where it formed. This is due

to the different magma formation processes, such as decompression-related partial melting (e.g., mid-ocean ridge), melting due to increased amount of volatiles (e.g., ocean–continent subduction), and melting due to increased temperature (e.g., continental collision). These various processes and the diverse melting materials will affect differently the chemical elements, which will be reflected in the chemical composition of the resulting igneous rocks.

3.4.1 Tectonic Provenance

We noticed earlier that the major element chemical composition of a rock reflects, via the major geochemical series, the tectonic setting where it formed. However, trace elements, or sometimes even a combination of major and trace elements, have been empirically demonstrated to be a better indicator of tectonic provenance. Let us consider a few examples.

If we consider only mafic rocks, a variety of discrimination diagrams is available (● Fig. 3.8). These use major elements (e.g., Si, Ti, Mn, Na), minor or trace elements (e.g., Y, Zr, Sr, Nb, P), or element ratios (e.g., Ti/Y, Nb/Y, Ce/Sr). Regardless of the elements used, they all provide similar information: a clear distinction between basalts from different tectonic settings. The most direct and useful information is the separation of ocean island basalt (OIB) originating from hot spots and thus representative of the deeper mantle, island arc basalt (IAB), produced at ocean–ocean subduction by partial melting due to the presence of volatiles, and mid-ocean ridge basalt (MORB), produced by partial melting of the upper mantle, due to decompression (● Fig. 3.8. Other diagrams add more fields, such as within-plate basalt (WPB) and calc-alkaline basalt (CAB), introduce more detailed petrological distinctions (e.g., ocean island tholeiite vs. ocean island alkaline basalt; ● Fig. 3.8).

A very similar approach can be taken for more felsic rocks, such as granites: several discrimination diagrams have been empirically devised, using mostly trace elements (e.g., Ta, Nb, Y, Rb), which fairly well discriminate between granitoids from different geodynamic environments (● Fig. 3.9). The main types of granites have been defined as syn-collisional granites(continent–continent collision), ocean ridge granites (normal and back-arc ridges), within-plate granites (rifting; also includes ocean islands), and volcanic arc granites (subduction-related; Pierce 1984).

Another rather interesting way to discriminate between the geodynamic contexts of granites is their oxygen isotope composition. From the point of view of oxygen isotopes, granites can be separated in three groups [39]:

— Low ^{18}O granites ($\delta^{18}O < 6‰$), found in ocean island arcs. Their formation has no contribution from continental crust, and they are thus purely mantle-derived.

◻ Fig. 3.7 A Harker diagram for a suite of igneous rocks, representing typical major element compositional trends, interpreted as reflecting magmatic differentiation and evolution

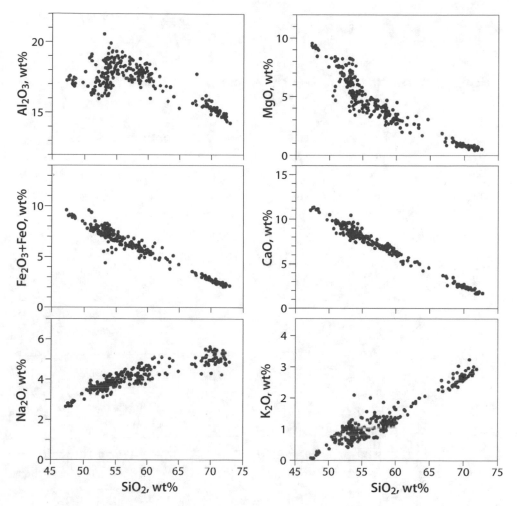

– "Normal" granites ($6‰ < \delta^{18}O < 10‰$), which tend to be mostly Precambrian. They derive from the partial melting of continental crust containing both volcanic and sedimentary components.

– High ^{18}O granites ($\delta^{18}O > 10‰$), which tend to be mostly Phanerozoic. They are likely derived from $\delta^{18}O$-enriched sediments and meta-sediments.

It is a little bit more difficult to derive the geodynamic context of basalts from their oxygen isotope composition, but there is some difference between the ocean basalts (mid-ocean ridge basalts, ocean island basalts, and island arc basalts), which have the have the lowest variability ($\delta^{18}O$ typically ranging from 5 to 7‰), and continental basalts (continental intraplate basalt, continental arc basalt, and continental flood basalt), which have a higher variability and a little bit higher values ($\delta^{18}O$ typically ranging from 4 to 10‰; [16]). These variations and differences in $\delta^{18}O$ values hint at the heterogeneity of the mantle, which we will discuss a little bit later in this chapter.

Following the same technique, the tectonic setting of sediments—in particular clastic sediments—can be derived, in addition to the degree of differentiation of the source rock (mafic, intermediate, felsic). In both cases, the major elements are considered more often than the trace elements. We can use the major elements' oxides directly, or we can recalculate them into some empirically derived "discriminant functions" ([4], ◻ Fig. 3.10). In either case, we can have a very clear idea of the geodynamic context where the sediments formed. This invaluable information can be useful for any reconstruction or regional geology study.

3.4.2 Mantle Heterogeneity

We mentioned earlier that the varying oxygen isotope composition of basalts suggests that Earth's mantle is not homogeneous. Logically, it would not be homogeneous, due to Earth's active tectonics: several processes contribute to the mantle's heterogeneity. For instance, the upper mantle is depleted of its most incompatible elements through the partial melting that generates mafic magmas at the mid-ocean ridge. The subducting oceanic crust, on the other hand, reintroduces more volatiles-rich mafic, and to a smaller extent felsic, material back into the mantle. The geochemical signatures of the different mantle reservoirs have been defined, using predominantly radiogenic isotopes, as we will see briefly now.

□ Table 3.1 Typical average chemical compositions of the C1 chondrite, selected main Earth reservoirs, and main rock types as function of differentiation

Element	Meteorites	Major Earth reservoirs			Main rock types			
	C1 chondrite	Bulk Silicate Earth	Upper mantle	Continental crust	Ultramafic	Mafic	Intermediate	Felsic
Major elements (wt%)								
SiO$_2$	0.86	44.92	44.92	58.19	42.35	50.27	65.26	69.45
TiO$_2$	0.004	0.20	0.15	0.48	0.05	0.50	0.56	0.43
Al$_2$O$_3$	0.86	4.44	4.21	15.83	2.27	15.64	15.7	14.73
FeO	18.1	8.05	8.05	7.84	12.40	11.06	4.28	3.81
MnO	0.019	0.13	0.13	0.10	0.20	0.23	0.08	0.07
MgO	9.65	38.13	37.80	4.85	38.47	7.54	2.02	1.83
K$_2$O	0.006	0.03	0.00	1.49	0.02	1.00	2.55	2.99
Na$_2$O	0.51	0.36	0.29	3.15	0.66	2.52	4.08	3.56
CaO	0.925	3.54	3.36	6.90	2.24	10.07	3.85	3.34
Lanthanides (ppm)					98.65	98.83	98.38	100.21
La	0.237	0.648	0.08	17	1.3	6.1	32.5	35.96
Ce	0.613	1.675	0.538	37.5	3.5	16	64.6	67.9
Pr	0.093	0.254	0.114	4.45	0.49	2.7	4.1	5.1
Nd	0.457	1.250	0.738	18	1.9	14	24.4	30
Sm	0.148	0.406	0.305	3.7	0.42	4.3	4.23	5.7
Eu	0.056	0.154	0.119	1.15	0.14	1.5	1.2	1.1
Gd	0.199	0.544	0.43	3.45	0.54	6.2	3.3	4.7
Tb	0.036	0.099	0.08	0.33	0.12	1.1	0.56	0.56
Dy	0.246	0.674	0.559	3.6	0.77	5.9	1.2	3.4
Ho	0.055	0.149	0.127	0.77	0.12	1.4	1.35	1.3
Er	0.16	0.438	0.381	2.2	0.3	3.6	0.6	2.1
Tm	0.025	0.068	0.06	0.32	0.041	0.6	0.5	0.45
Yb	0.161	0.441	0.392	2.1	0.38	3.2	1.4	2.25
Lu	0.025	0.068	0.061	0.32	0.036	0.55	0.23	0.35
Other elements (ppm)								
Ag	0.2	0.008	0.000784	0.08	0.05	0.11	0.052	0.04
As	1.85	0.05	0.01	1	0.8	2.2	1.3	1.7
Au	0.14	0.00098	0.00096	0.003	0.006	0.004	0.0007	0.004
B	0.9	0.3	0.075	10	2	5	5.8	9.5
Ba	0.00241	0.0066	0.449	320	0.7	315	837	614
Be	0.025	0.068	0.0442	1.5	0.2	0.7	1.6	2.5
Bi	0.011	0.0025	0.0005	0.06			0.16	
Br	3.57	0.05	0.005		0.8	3.3		1.3
C	35,000	120	18					
Cd	0.71	0.04	0.0392	0.098	0.05	0.21	71	0.13
Cl	680	17	2.55		45	55	108	180
Co	500	105	105	27	175	47	13.5	17.3
Cr	2650	9000	2625	152	1800	185	38.3	56.2
Cs	0.19	0.021	0.504	1.8	0.1	1.1	2.3	3

◻ **Table 3.1**   (continued)

Element	Meteorites	Major Earth reservoirs			Main rock types			
	C1 chondrite	Bulk Silicate Earth	Upper mantle	Continental crust	Ultramafic	Mafic	Intermediate	Felsic
Cu	120	30	29.1	49.5	15	94	22.7	18.1
F	60	15	9.75		100	520	570	850
Ga	9.2	4	3.8	17	1.8	18	19.9	20.2
Ge	31	1.1	1.1	1.6	1.3	1.4	1	1.3
Hf	0.103	0.283	0.167	3.35	0.4	1.5	4.8	4.7
Hg	0.3	0.01	0.0098		0.01	0.09	8.7	0.08
I	450	0.01	1		0.3	0.5		0.5
In	0.08	0.01	0.0085	0.05	0.012	0.22		
Ir	0.455	0.003185	0.003185	0.0001				
Li	1.5	1.6	1.568	12	0.5	16	15	36
Mo	0.9	0.05	0.03	1	0.3	1.5	0.82	1.2
Nb	0.24	0.658	0.11186	11.5	9	20	11.3	10.3
Ni	10,500	1960	1960	78	2000	145	19.4	22.8
Os	0.49	3.43	3.43					
P	1080	90	54	873	195	1130	742	0.12
Pb	0.00247	0.00015	0.000018	10.3	0.5	7	18.2	23.9
Pd	0.55	0.00385	0.000385	0.001			0.0008	
Pt	0.101	0.00707	0.00707				0.00093	
Rb	2.3	0.6	0.0408	45	1.1	38	72.4	109.2
Re	40	0.28	0.27	0.0005				
Rh	0.13	0.00091	0.00091					
Ru	0.71	0.00497	0.00497					
S	54,000	250	238		200	300		300
Sb	0.14	0.0055	0.0011	0.2	0.1	0.6	0.21	0.2
Sc	5.92	16.2	15.39	26	10	27	10.4	12.3
Se	21	0.075	0.07125	0.05	0.05	0.05	0.0577	0.05
Sn	1.65	0.13	0.0975	2.5	0.5	1.5	1.2	3
Sr	7.25	19.865	12.935	292.5	5.5	452	490	296.4
Ta	0.014	0.037	0.006	1.05	0.5	0.8	0.81	0.9
Te	2.33	0.012	0.01176					
Th	0.029	0.079	0.006	4.55	0.0045	3.5	8.8	11.4
Tl	0.14	0.0035	0.00035	0.91	0.04	0.21	0.38	2.3
U	0.0074	0.02	0.002	1.17	0.002	0.75	1.9	2.48
V	56	82	82	180.5	40	225	67	66.1
W	0.093	0.029	0.002	1	0.5	0.9	0.4	2
Y	1.57	4.302	3.655	20		21	16.3	25
Zn	310	55	55	76.5	40	118	70.7	63.6
Zr	3.82	10.467	6.195	111.5	38	120	177	163.6

Data from Turekian and Wedepohl [41], Vinogradov [42], Taylor and McLennan [40], and Faure [14]; data within Gao et al. [15], Kemp and Hawkesworth [20], Walter [43], and Albarede [1]; and data within and earthref.org

3

◻ **Fig. 3.8** A selection of
basalt discrimination diagrams,
clearly differentiating them by
geodynamic context, but also
by belonging to a major igneous
rock series (e.g., tholeiitic vs.
alkaline). Abbreviations: MORB:
mid-ocean ridge basalt, OIB:
ocean island basalt, IAB: island
arc basalt, CAB: calc-alkaline
basalt, IAT: island arc tholeiite,
OIT: ocean island tholeiite,
OIAB: ocean island alkaline
basalt. From Mullen [25], Pearce
and Cann [31], Pearce and Norry
[28], Pearce [29], and Pearce [30],
modified

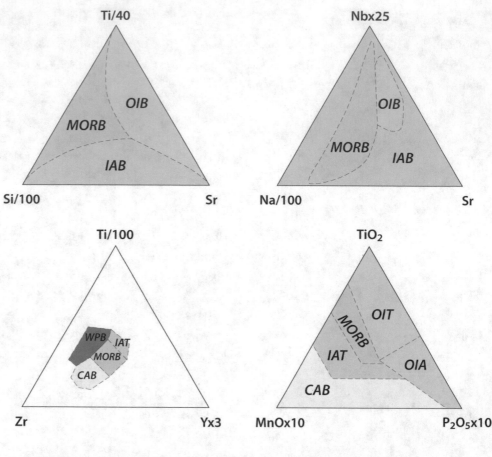

◻ **Fig. 3.8** A selection of
basalt discrimination diagrams,
clearly differentiating them by
geodynamic context, but also
by belonging to a major igneous
rock series (e.g., tholeiitic vs.
alkaline). Abbreviations: MORB:
mid-ocean ridge basalt, OIB:
ocean island basalt, IAB: island
arc basalt, CAB: calc-alkaline
basalt, IAT: island arc tholeiite,
OIT: ocean island tholeiite,
OIAB: ocean island alkaline
basalt. From Mullen [25], Pearce
and Cann [31], Pearce and Norry
[28], Pearce [29], and Pearce [30],
modified

◻ **Fig. 3.9** Examples of
chemical discrimination
diagrams, using trace
elements, defining granitoids
by their geodynamic context.
Abbreviations: SCG: syn-
collisional granite, ORG: ocean
ridge granite, WPG: within-
plate granite, VAG: volcanic
arc granite. After Pierce (1984),
modified

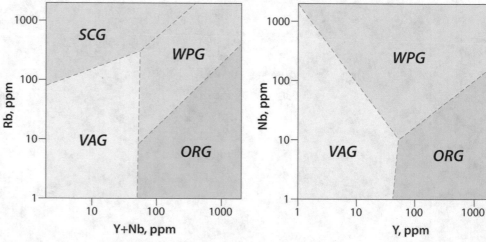

Rb is a highly incompatible element, enriched in highly differentiated rocks; it also readily substitutes for K (see ◻ Table 1.1). It is also radioactive and disintegrates to Sr (on which the $^{87}Rb/^{86}Sr$ geochronological method is based). Because of rubidium's high incompatibility, we will use it as a proxy for the extent of differentiation of an igneous rock; we will use the ratio $^{87}Sr/^{86}Sr$, where ^{87}Sr is proxy for ^{87}Rb and ^{86}Sr is stable and non-radiogenic. Therefore, higher the $^{87}Sr/^{86}Sr$ ratio in a rock, the higher its degree of differentiation is. There is another isotope system that can also be used for that

purpose, ^{147}Sm-^{143}Nd, with one crucial difference: Sm is highly compatible, preferentially fractionating into the crystalline phases during igneous processes. Thus, high $^{143}Nd/^{144}Nd$ ratios will correspond to lower degrees of differentiation. We can plot the two isotope systems against each other and derive a general differentiation trend for Earth's mantle (◻ Fig. 3.11).

We can use other isotopic systems as well, such as Re-Os or U–Pb, and the implications would be the similar. However, things are a bit more complicated when Pb isotopes are considered (◻ Fig. 3.12): apart from

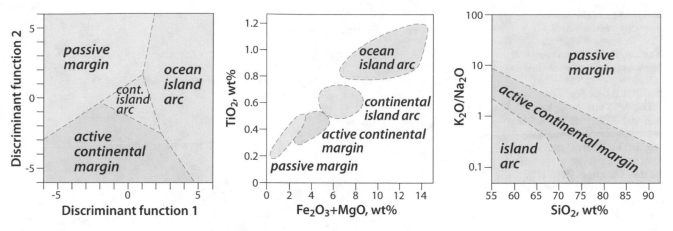

Fig. 3.10 Examples of discriminant diagrams for sedimentary rocks: sandstones (left and middle) and sandstone–mudstone suites (right), based on the major element compositions and clearly showing the geodynamic context where the sediments formed. From Bhatia [4] and Roser and Korsch [34]. The "discriminant functions" are based on all the major elements

Fig. 3.11 Nd and Sr isotope composition plot for oceanic basalts, with two main compositional trends, one corresponding to magma fractionation (or differentiation) and the other to the recycling of sediments into the mantle. The depleted and the enriched quadrants are defined relative to the Bulk Silicate Earth (BSE). From Dosso and Murthy [10] and Hofmann and White [19], modified

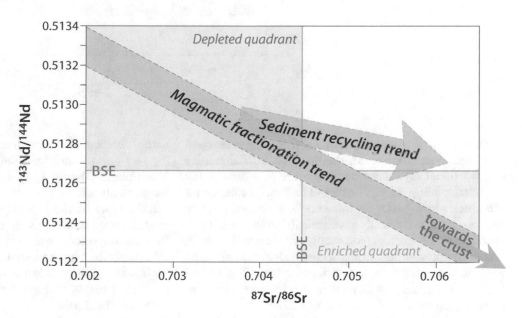

the major ones, the mid-ocean ridge basalt (MORB) and the ocean island basalt (OIB, originating from hot spots), two other reservoirs are apparent: one with high $^{87}Sr/^{86}Sr$ ratios, called the enriched mantle, and another, with high $^{206}Pb/^{204}Pb$ ratios, called high μ. The μ value is equal to the $^{238}U/^{204}Pb$ ratio and varies from around 7 for the mantle to around 10 for the highly differentiated upper crust; this means that high μ values correspond to a high degree of mantle differentiation.

It is somewhat unclear at present exactly what the origins of the enriched mantle and high μ are. Given that high μ values correspond to differentiated rocks from the upper continental crust, one possible interpretation is that the high μ reservoir corresponds to such rocks that have been incorporated—through subduction and continental collision—back into the mantle. It is possible that a similar explanation can also be applied to the enriched mantle, even though it has very different Sr and Pb ratios.

3.4.3 Magma Differentiation, Source of Sediments, and Degree of Alteration

As we discussed earlier, the extent of differentiation of an igneous rock is clearly reflected in its chemical composition—for both major and trace elements—and can be derived from that. There are two geochemical methods that are specifically designed to study the differentiation of igneous rocks: REE spectra and Sr and Nd isotopes; we will discuss these here.

3.4.4 Magma Differentiation Through Partial Melting and Fractional Crystallization

Broadly speaking, there are 6 types of mechanisms resulting in magma differentiation (partial melting, fractional crystallization, magma assimilation and magma mixing, magma flow segregation, transport of volatiles, and crustal

3

◧ **Fig. 3.12** Main mantle
reservoirs, represented in a
$^{87}Sr/^{86}Sr$ versus $^{206}Pb/^{204}Pb$
diagram. Three distinct reservoirs
become apparent: the enriched
mantle (high $^{87}Sr/^{86}Sr$ ratios),
the MORB (low $^{87}Sr/^{86}Sr$ and
$^{206}Pb/^{204}Pb$), and the high μ,
with high $^{206}Pb/^{204}Pb$ ratios. The
ocean island basalts—normally
originations from hot spots—
seem to be a mixture of the three
sources. Data from White [45],
Zindler and Hart [47], and Hauri
et al. [17]

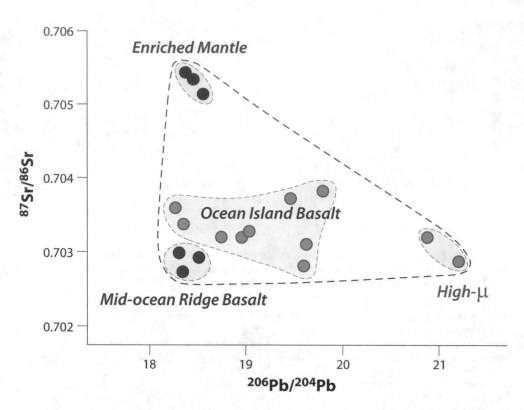

assimilation and contamination). Of these, the most common and most effective are the first two, partial melting and fractional crystallization, which are in a certain way the mirror image of each other. The first begins with a solid rock which, under the influence of increased pressure and temperature (typically in continental collision setting), introduction of volatiles (commonly seen in a subduction), and decompression (as in mid-ocean ridge setting), starts to melt and generate a silicate liquid phase. The second occurs in a magma chamber where, due to decrease in temperature, the silicate melt starts to crystallize.

As a result, in those two cases, partial melting and fractional crystallization, we have the *coexistence*, commonly under equilibrium conditions, of a silicate liquid phase and a crystalline solid phase. As we will recall from Chap. 1, ▶ Sect. 1.6.2, in such a situation some elements will prefer to segregate in the liquid phase (the incompatible ones) and some elements preferentially segregate in the solid phase (the compatible ones). It is this specific behaviour of the elements that will condition, indeed, the process of magma differentiation and the generation of different types of magmas and therefore of different igneous rocks. These are well-understood processes that are at the core of our igneous petrology knowledge and studies.

We can go one step further than these general principles and consider the extent of magma differentiation that will occur as function of how much of the initial rock has been melted (in the case of partial melting) or how much of the initial magma has crystallized (in the situation of fractional crystallization).

In the first case, that of partial melting, *the highest degree of differentiation will occur at the lowest degree of*

partial melting: the most incompatible elements will be the very first to leave the solid phase and segregate in the liquid phase (◧ Fig. 3.13). Inversely, the most compatible elements will be the last to leave the solid phase and will therefore be the last to go into the liquid phase. Of course, once the entirety of the initial rock has melted, the resulting silicate melt will have the exact same chemical composition as the initial rock (◧ Fig. 3.13).

Inversely, in the situation of fractional crystallization, the first elements that will be extracted from the silicate melt—in the form of a crystalline cumulate—will be the most compatible ones (◧ Fig. 3.13). In this case, *the highest degree of magma differentiation will be achieved at the highest degree of crystallization*, when only the most incompatible elements will stay in the remaining silicate melt (◧ Fig. 3.13).

When silicate melt is generated during partial melting, the new silicate melt has lower density than the initial rock and will most likely migrate upwards, thus removing the most incompatible elements away from the melting site and upwards in the crust, leaving behind the more compatible elements. In a similar way, during fractional crystallization the most incompatible elements will stay in the remaining fluid which will become less dense and less viscous (as enriched in volatiles) and will also tend to move upwards in the crust, away from the initial magma chamber and possibly towards a new one. These two processes, partial melting and fractional crystallization, will often occur not in a single stage, or step, but repeatedly over the thickness of the crust. In other words, higher degree of differentiation will be achieved in a thicker crust than in a thinner one: for instance, only one "degree" of dif-

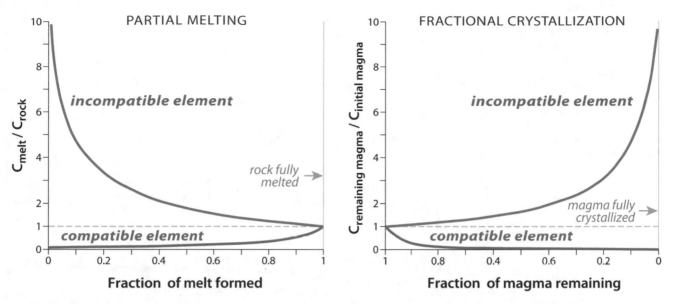

□ Fig. 3.13 General form of the relationship between extent of differentiation and the degree of partial melting (left) and the degree of crystallization (in the case of fractional crystallization, right). In the case of partial melting, incompatible elements will be the most enriched in the newly generated silicate melt at the lowest degree of melting. In the case of fractional crystallization, the incompatible elements will be the most enriched in the remaining magma at the highest degree of crystallization

ferentiation will occur at mid-ocean ridge settings (from ultramafic to mafic lithologies), as only one step of partial melting and one stage of fractional crystallization occur. Inversely, several degrees of magma differentiation are present on a continental collision setting, where partial melting can occur more than once at the bottom of the crust and fractional crystallization will occur in a succession of magma chambers within the thickened crust.

3.4.5 Rare Earth Elements as Proxy for Magma Differentiation

Rare Earth Elements (REE), or lanthanides, are transition metals from the third group of the periodic table (□ Fig. 1.3). They all have the same valance, +3 (with the exception of Ce, also present as Ce^{4+} and Eu, also present as Eu^{2+}), and very similar chemical properties and geochemical behaviour, due to their particular electronic configuration. They differ from each other in one particular respect: their ionic radii, which decrease from approximately 1.1 Å for La to approximately 0.9 Å for Lu (□ Table 1.1). This phenomenon is known as the lanthanide contraction. Because of this, and keeping in mind that they have—mostly—the same valance state, their ionic potential and thus their *compatibility* will increase from La to Lu: the light REE (LREE, let's say La to Nd) will be relatively incompatible, whereas the heavy ones (HREE; typically Dy to Lu) will be relatively compatible. The result of this different geochemical behaviour is that the LREE will be significantly enriched in highly differentiated rocks (e.g., pegmatites, leucogranites, carbonatites) and the HREE will be depleted in these rocks.

We usually plot REE compared to the C1 chondrite (values in □ Table 3.1; □ Fig. 3.14), which we consider to represent the average bulk chemical composition of the solar system, and thus the least differentiated REE compositions. The diagrams thus obtained (□ Fig. 3.14) are called REE patterns, or *spectra,* and are very widely used. Indeed, they are the second plot we might want to generate for an igneous rock that we are studying, after TAS, assuming the REE data are available. In general terms, it is interpreted with regard to both the absolute values and the shape of the patterns:

- For ultramafic to mafic rocks, the C1-normalized values will be typically under 10 and will tend to be depleted in LREE (pattern gently sloping to the left).
- For intermediate to felsic rocks, the *C*1-normalized values will typically be above 10, often above 100. The pattern will slope to the right, with higher degrees of differentiation, and will reflect in steeper slopes, strong Eu anomaly, and very low HREE values.

The Eu anomaly mentioned earlier is due to magmatic processes occurring in a magma chamber, specifically fractional crystallization. As Eu has a second valance state, Eu^{2+}, it will readily substitute for Ca in anorthite. Ca is a compatible element that will form minerals and thus leave the silicate melt at higher temperature. These minerals will settle to form the cumulate at the bottom of the magma chamber. If a rock is formed from the remaining silicate melt (either higher in the same magma chamber or, more likely, in a separate magma chamber), it will be already depleted in Eu, which will be clearly visible in its REE pattern. Rocks that have formed after repeated episodes of fractional crystallization are

3

◻ Fig. 3.14 Typical *C*l chondrite-normalized Rare Earth Elements patterns, or spectra, for some main Earth reservoirs: depleted upper mantle, MORB (proxy for the average oceanic crust), and average continental crust; an example of highly fractionated plutonic rock (leucogranite). Data from White [46] and Rudnick and Fountain [35]

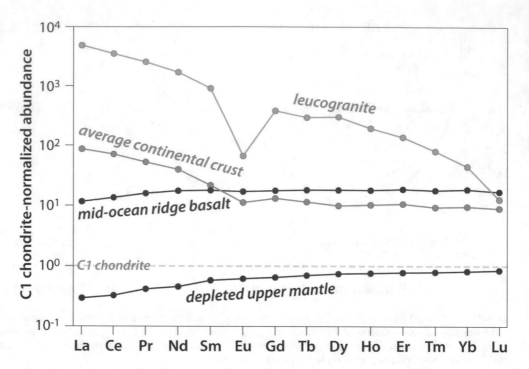

necessarily very depleted in Eu, as it "stayed behind" in the solid phase at each such episode. This is why strong Eu anomalies are interpreted to indicate high degrees of magma differentiation (◻ Fig. 3.15).

As a corollary, a REE pattern with positive Eu anomaly will be observed for a rock corresponding to the cumulate left behind in the magma chamber and enriched in more compatible elements. Given that the extent of Eu anomaly (denoted Eu*) is clearly indicative of the extent of magma differentiation, it is useful to calculate it, which is done by simply comparing it to the *Eu* value that would be expected without any anomaly:

$$Eu^* = \frac{Eu_{measured}}{Eu_{expected}} = \frac{Eu_{measured}}{(Sm_{measured} + Gd_{measured})/2}$$

Thus, the Rare Earth Elements are a very useful tool for studying igneous processes, as their only geochemical behaviour difference—their degree of compatibility—allows us to visualize and even measure the extent of magma differentiation, not only during fractional crystallization, but also during partial melting where the exact same principles apply, only in reverse.

There are other ways to measure the extent of differentiation, using radiogenic isotopes. We already mentioned that Rb, a highly incompatible element, disintegrates to Sr ([87]Rb → [87]Sr); this is the base of the Rb–Sr geochronological method that we will discuss in more detail later in this chapter. We can use uniquely the [87]Sr/[86]Sr ratio, where [87]Sr is proxy for [87]Rb and [86]Sr is stable and non-radiogenic. Therefore (as mentioned earlier), the higher the [87]Sr/[86]Sr ratio in a rock, the higher is its degree of differentiation. On top of this, the age of an igneous rock will also affect its [87]Sr/[86]Sr ratio. As a result, each major Earth reservoir, but also each particular igneous (or metamorphic)

rock will have its own and specific [87]Sr/[86]Sr ratio, which can be very useful for a variety of endeavours.

3.4.6 Rare Earth Elements in Sedimentary Rocks

Another application of the Rare Earth Elements (REE, or the lanthanides) is to study sedimentary rocks and specifically their provenance and the processes that affected them during and after their formation. Generally speaking, there are two mobilization and transportation modes for REE: (1) as mineral particles during physical weathering and clastic transport and sedimentation, and (2) as dissolved ions in river water and seawater.

The REE composition of clastic rocks, in the first scenario, will be strongly affected by the source of sediments—their *provenance*—whereas other factors (e.g., processes such as weathering and diagenesis) have negligible overall effect. Only in the case of extreme weathering will we see significant differences between the REE composition of the source rocks and the resulting clastic sedimentary rocks. As a result, we can, and often do, use the REE signature of a clastic rock to study its provenance, in a way not dissimilar to finding a sedimentary rock's geotectonic setting that we examined earlier (at the end of ▶ Sect. 3.4.1; ◻ Fig. 3.16), which can also be very advantageous in paleo-reconstructions. For instance, a progressive increase in the Light Rare Earth Elements (LREE) from oceanic island arc to continental island arc to active continental margin to passive margin settings was noted [5], which is likely explained by the progressively more differentiated igneous rocks present in these settings. Additionally, a relationship between the maturity of clastic sediments and

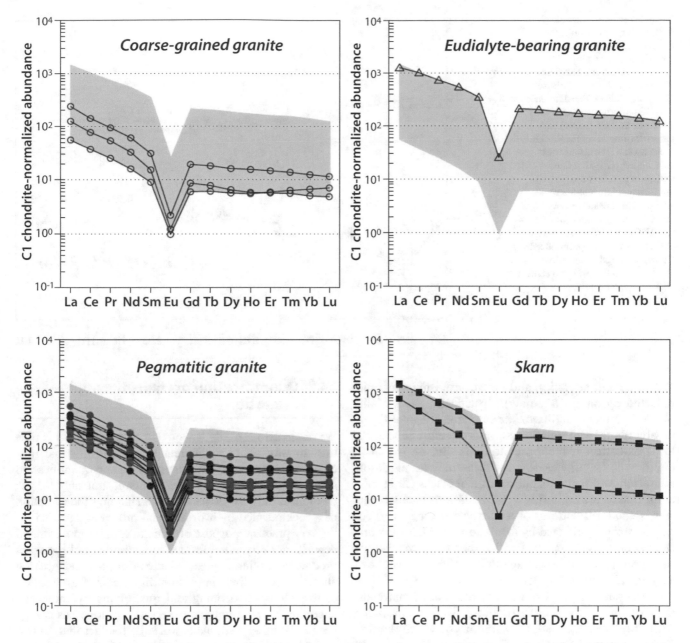

Fig. 3.15 Example of REE patterns for felsic rocks (three peralkaline granites and a skarn) from the Ambohimirahavavy alkaline complex, Madagascar. The grey area contours the entire compositional range of the studied rocks. Both the high values of C1 chondrite-normalized REE concentrations (typically above 100 for the LREE) and the pronounced Eu anomaly clearly indicate fairly highly differentiated rocks. After Estrade et al. [13], modified

their overall REE abundance and Eu anomaly was noted [5], suggesting that diagenesis of clastic rocks will have some effect on their REE chemistry. Other factors that must also be considered when using the REE chemistry of a clastic rock to study its provenance are the grain size and sorting, and mineralogical composition of the sediment. The finer grain size fractions represent the REE of the source more faithfully (and also have higher abundances of REE). As a corollary, it may be preferable, in particular in cases of high "dilution" by quartz and carbonates, to extract and study only the clay size fraction of the clastic rock studied. On the other hand, the presence of heavy minerals rich in REE (e.g., zircon and monazite) may

increase the overall REE abundance of the clastic rock, but introduce significant heterogeneity, as these minerals are often not uniformly distributed in sedimentary rocks.

The situation with chemical sediments is very different: their REE compositions generally tend to reflect the REE abundances of seawater, which are very low, are heavy REE-dominated, and have a pronounced negative Ce anomaly [12]. This is often observed in ferromanganese nodules from the seafloor [11] or in limestones (□ Fig. 3.16), as well as in Mn and Fe sedimentary deposits. However, the general tendency of sediments "inheriting" the seawater REE pattern is often disrupted by local redox conditions (reflected by the Ce anomaly)

3

Fig. 3.16 Examples of REE patterns of sedimentary rocks (normalized to the North American Shale Composite, or NASC) from the Youjiang–Nanpanjiang Basin (south-west Guizhou, China). The difference in REE patterns between clastic (siltstone and sandstone) and biogenic (limestones) sedimentary rocks is clear: the clastic rocks have higher REE abundances and seem to reflect the source (an intermediate igneous rock), whereas the limestones have much lower REE abundances (as the REE abundances of seawater are very low) and a pronounced Ce anomaly, reflecting redox conditions. Data from Bao et al. [2]

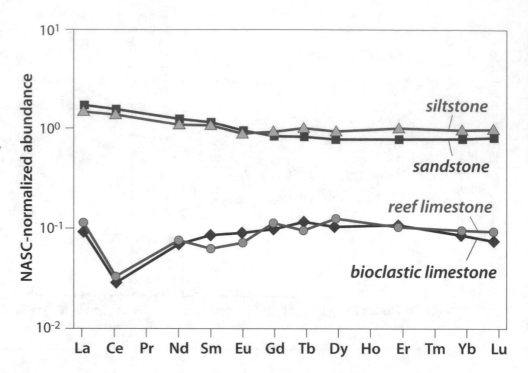

or subsequent alteration and metamorphism (sometimes reflected by an Eu anomaly). Specifically, the mobility of Ce is affected by its oxidation state (+3 or +4) and its abundance will be affected, and therefore reflect, the local oxidation–reduction conditions. Eu, on the other hand, will be affected by hydrothermal input during alteration and metamorphism, as it replaces Ca in minerals, due to it also having a +2 oxidation state. In other words, the REE patterns of a chemical sedimentary rock will be strongly affected by formation and alteration *processes* and the *conditions* at which they that occurred; we can use this to our advantage to decipher the history of a chemical sedimentary rock (e.g., Su et al. [38]).

REE patterns for sedimentary rocks are constructed in a manner very similar to those for igneous rocks, with only one difference, the standard by which the concentrations are normalized. For sedimentary rocks, the general idea is to normalize relative to the *upper continental crust*, from where the source material presumably comes. Commonly used standards include the North American Shale Composite (NASC, as in the example in Fig. 3.16), the Post-Archean Australian Shale (PAAS), and the Upper Continental Crust (UCC). However, the standard we choose will depend on the specific problematics or areas studied and can be pretty much anything, with sometimes even chondrites being used.

In conclusion, REE patterns and abundances of sedimentary rocks can be invaluable in studying their *provenance* (for clastic rocks) and the *processes* that affected them (for chemical rocks). This is particularly valuable when the origin and subsequent alteration and metamorphism of Mn and Fe sedimentary deposits are studied, including banded iron formations.

3.4.7 Quantification of Hydrothermal Alteration

We discussed primarily igneous processes so far. It is now time to turn on to a different topic, hydrothermal processes and, specifically, the qualification and quantification of alteration. It is a very useful pursuit, not only for the theoretical purpose of better understanding the evolution of a rock affected by hydrothermal processes, but also for the very practical purpose of mineral exploration. Mineral deposits are very often formed by hydrothermal processes producing varying degrees of alteration, and quantifying alteration has a direct impact on the success of exploration efforts. However, defining and quantifying hydrothermal alteration by petrographic observations alone has always been a daunting task, made worse by the fact that no two geologists will agree, based purely on visual observation, the degree of alteration of a rock. We will see here how geochemistry can help with both the qualification and the quantification of hydrothermal alteration.

There are three main methods to qualify and quantify hydrothermal alteration, and we will consider these in turn: normative mineralogy, isocons, and Pearce element ratios (PER).

We mentioned briefly normative mineralogy as a way to classify igneous rocks as function of their bulk major element compositions. Here, we will do something very similar: we will calculate, on the basis of the rock's bulk chemical composition, what alteration minerals may be present and in what amounts. Let's illustrate this approach with a simple example.

Let us imagine a fairly mature sandstone basin, in which are present zones of intense clay alteration related to mineral deposits. We know that the clay minerals are illite $(KAl(Si_{4-x}Al_x)O_{10}(OH)_2)$, chlorite $((Mg_{5-x}Fe_xAl)(Si_{4-x}Al_x)O_{10}(OH)_2)$, and kaolinite $(Al_2Si_2O_5(OH)_4)$. On the basis of this knowledge, we will assign all K from the bulk geochemistry to illite; we will also assign the corresponding amounts of Si and Al, based on the structural formula proportions. Then, we will assign all Fe and Mg from the bulk geochemistry to chlorite, as well as the corresponding Si and Al. Finally, we will assign the remaining Al from the bulk rock geochemistry to kaolinite, together with the corresponding Si. (As a precaution, we might want to assign the remaining Si to quartz and calculate its amount, too: the sum of the four minerals should add up to 100%.) These calculations will provide us with the weight percentage amounts of the three alteration minerals; knowing their density we can also calculate their volumetric percentage. The calculations have several advantages: they are relatively easy and straightforward, they can be automated and be done by a computer for a vast number of samples, they can be applied to a variety of situations, and they are valid even with relatively complex alteration patterns. Indeed, it is in complex situations that their usefulness is most apparent. This approach requires that we have some background knowledge of the local geology and of the alterations present, but this is the case with all geochemical methods: geochemistry does not work in isolation but by integration with all available knowledge of the local and regional geology.

The second method we will consider consists of plotting, against each other, the average chemical compositions of two different rock units. This can be the unaltered and altered varieties of the same rock—with each unit on one of the axes—but it can also be used for any two rocks. This is called the isocon method and is very practical and useful (■ Fig. 3.17), as it allows appreciating, at a glance, which elements are predominantly enriched in which lithology or unit.

This method is primarily visual and as such allows for a very quick identification and interpretation of chemical composition differences between two rock units. However, it can also have specific numerical information added to it (■ Fig. 3.18), making it also a quantification tool. Simple as it is—and its simplicity itself adds to its worth—it is a central geochemical visualization tool and should be used widely in any sort of investigation.

3.4.8 Pearce Element Ratios

A more rigorous alteration quantification method, the Pearce element ratios, was developed by the eponymous Canadian geologist T. H. Pearce [32]. It can also be applied to the quantification of igneous processes (e.g., magma differentiation) or any other problem involving the geochemical comparison of two lithologies. In our case, we will focus on an alteration example, remembering that the same principles will apply if we use the method in other situations.

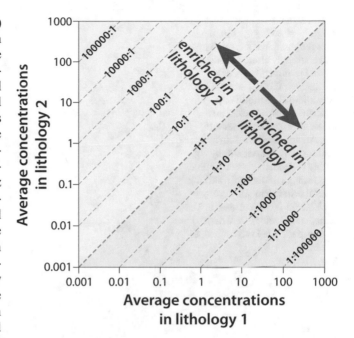

■ **Fig. 3.17** Graphic base for the isocon diagram, where the average composition of an element in one lithology is plotted against its average composition in another lithology. The actual average values (mean? median?) come from a simple statistical analysis of the samples and can be provided in a table form. The visual interpretation is straightforward: if the element sits on the 1:1 line, then there is no difference between its contents in the two lithologies; if it sits above the 1:1 line, it is enriched in lithology 2; if it sits below the 1:1 line, it is enriched in lithology 1. The two axes always use logarithmic scale, as this allows us to visualize many elements with extreme compositional differences—from trace to major—on the same diagram

The Pearce element ratios (PER) method relies on using element ratios rather than the elements in themselves. We always use molar ratios, as the interpretations will be in terms of mineralogical changes in the rock. The general format is to normalize the elements that we expect will change—whose change we want to study—to some other elements that we expect will remain rather immobile (or "*conservative*", meaning neither added nor removed during alteration).

Thus the first step in the process is to verify if there are conservative elements in the lithologies we study. We plot the most likely candidates (e.g., Zr, Ti, Al, Nb) against each other (■ Fig. 3.19) and can visualize how well they correlate. If they do reasonably well and the correlation regression goes through the origin, then we can conclude that they are preserved during the alteration process and can be reliably used as a common denominator. However, if they are not preserved due to fairly strong alteration (■ Fig. 3.18), we can use the most conserved element as a common denominator (often Al). This version of the method is called general element ratios (GER) and gives similar and similarly reliable results [37].

The second step is to design an appropriate diagram—and the related calculations—that are applica-

3

3

Fig. 3.18 Example of isocon diagram, showing the difference between unaltered sandstone and a strongly altered sandstone related to mineral deposits (Athabasca Basin, Canada; from Alexandre 2020 modified). The four most enriched elements in the altered sandstones—Bi, Co, As, and Cr—are clearly visible (and are accidentally highly to moderately compatible, suggesting a mafic contribution); several other elements have enrichment factors of 5—10. The enrichment factors were derived by comparing the median values

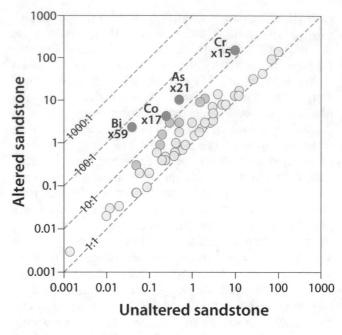

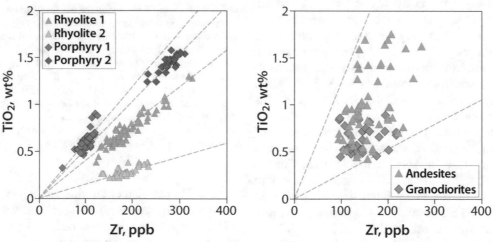

Fig. 3.19 Two rather immobile elements, Ti and Zr in this case, plotted against each other to verify if "conservative" elements are present in the lithologies studied, so that they can be used in Pearce element ratios. On the left diagram, Zr and TiO₂ correlate reasonably well—on regression lines going through the origin—suggesting that they have not been mobilized to any significant degree and thus can be used to quantify alteration. On the right diagram, Zr and TiO₂ do not correlate at all, likely due to a very advanced alteration, and cannot be used in Pearce element ratios; in this case the most conserved element will be used

ble in the specific case. Let us take as an example the Elura Zn-Pb-Ag deposit in New South Wales, Australia (■ Fig. 3.19), hosted by turbidites and associated with the subtle, even cryptic development of siderite and the destruction of albite [23]. The most valuable vector to ore would be the development of siderite, and it is this alteration that the method will attempt to detect and quantify. The ratios used will be Ca/Ti and C/Ti: less calcite ($CaCO_3$) or dolomite ($(Mg, Ca)CO_3$) available will be interpreted as more siderite ($FeCO_3$) present (■ Fig. 3.19).

Crucially, we can not only visually define and qualify the type of alteration (siderite in this case), but also quantify the degree of alteration for each sample. Using the appropriate ratios (e.g., Ca/Ti in this case) we can

calculate how far each sample is from the calcite and from the dolomite control lines and thus how altered it is: the farther away from the control lines a sample is, the more altered it is. In this manner—and after some reasonably complicated mathematical formulations unique for each specific case—we can define the type and degree of alteration for each individual sample. This is very valuable, in particular in cases of subtle alteration that are not easily detected by visual observation (■ Fig. 3.20).

Let us consider another example, that of muscovite and chlorite alteration associated with VMS deposits in the United Verde VHMS camp, Arizona, USA. Here, the host rocks are alkali feldspar-bearing rhyolites and dacites, and the alterations are not easily detected by visual observation. Two trends are present here

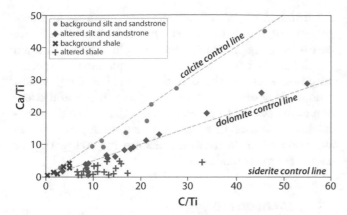

Fig. 3.20 Pearce element ratios diagram for carbonate minerals present at the Elura Zn-Pb-Ag deposit in Australia, with Ca and C normalized by the conservative Ti. The lines on which calcite, dolomite, and siderite would be situated are indicated. The samples situated below the dolomite control line will be enriched in Fe relative to the other samples and thus indicative of ore-related siderite alteration. After McQueen and Whitbread [23], modified

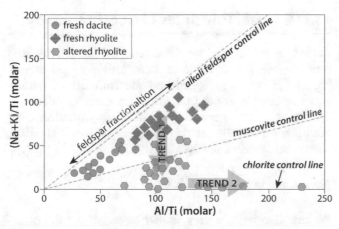

Fig. 3.21 An example of a PER diagram with Al/Ti plotted against (Na + K)/Ti, for whole rock samples associated with the VMS deposits of United Verde VHMS (Arizona, USA). The variation along the alkali feldspar control line is very likely due to varying degrees of magma differentiation (dacite has lower Na + K than rhyolite as it is less differentiated). Fresh dacite and rhyolite are situated below the alkali feldspar control line, indicating a varying degree of muscovite alteration; this is more visible with the altered rhyolites, which also fall below the muscovite control line, reflecting chlorite alteration (Trend 1). The increase in Al/Ti values along the chlorite control line likely reflects the formation of secondary chlorite (Trend 2). After Stanley [36], modified

(▪ Fig. 3.21): the first trend is vertical, from the alkali feldspar control line downward across the muscovite control line to the chlorite control line, and reflects the effects of the feldspar-to-muscovite and muscovite-to-chlorite alteration reactions; the second trend is horizontal and very near the horizontal axis and therefore likely represents a second chlorite-forming reaction (▪ Fig. 3.21).

Just as with the carbonate mineral example we saw earlier, in this second example we can calculate the molar proportion—or weight proportion—of the alteration minerals present in any particular sample. To do so

we use the so-called *mixing calculations*, which help us calculate the chemical composition of a mixture on the basis of the chemical composition of the components involved and their relative proportions.

3.4.9 Mixing Calculations

The first and most straightforward case is calculating the chemical composition of a mixture for a single element or a simple binary mixture. We use the equation

$$[X]_M = [X]_A f_A + [X]_B f_B$$

where $[X]$ is the concentration of the element X in the mixture and in the participating components A and B and f_A and f_B and the relative proportions of the two components. Given that $f_A + f_B = 1$ and therefore $f_B = 1 - f_A$, we can rewrite the equation to

$$[X]_M = [X]_A f_A + [X]_B (1 - f_A)$$

We can apply the same idea to a mixture of three components—a ternary mixture—or even more:

$$[X]_M = [X]_A f_A + [X]_B f_B + [X]_C f_C$$

We will always assume, due to the concept of mass conservation, that $f_A + f_B + f_C = 1$.

Things will get a bit more complicated when we calculate element or isotope ratios (e.g., Na/Ti and ^{143}Nd/^{144}Nd) or isotopic compositions of one element (e.g., δ^{18}O) for a mixture. In that situation, we have to consider not only the ratios or isotopic compositions of the participating components and their relative proportions (f_A and f_B), but also another weighing factor, which is the relative amounts of the denominators (in the case of element ratios, Ti or ^{144}Nd in this example), or the relative concentrations of the participating elements (in the case of isotopic composition; O in this example).

In the case of an isotope mixture of one element, the mixing equation is the following (taking oxygen isotopes as an example):

$$\delta^{18}O_M = \delta^{18}O_A f_A ([O]_A / [O]_M) + \delta^{18}O_B f_B ([O]_B / [O]_M)$$

where $[O]$ is the concentration of O in the mixture and in the participating components A and B. The resulting mixing line is not a straight line but a hyperbole. Similar to single element mixing, we can reformulate the equation for the mixing of three or more components, always assuming that their relative proportions combine to 1. However, if the concentrations of the element in the participating sources are the same, then the mixing equation is the same as for a single element.

In the case of two different element or isotope ratios (e.g., Na/Ti versus Ca/Zr or ^{143}Nd/^{144}Nd versus ^{87}Sr/^{86}Sr) things start to become a little bit complicated. The ratios of the mixture will depend not only on the rations of the participating sources and their relative proportions, but also on a "*curvature factor*", commonly written as *r*, rep-

3

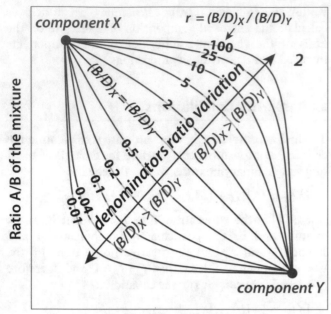

Ratio C/D of the mixture

☐ **Fig. 3.22** General shapes of the hyperbolic lines describing the mixing of two sources, *X* and *Y*, for two different elements' isotope ratios. The curvature factor r depends on the denominator ratios. After Langmuir et al. [21], modified

resenting the relative denominator ratios and controlling the curvature of the hyperbolic mixing line (☐ Fig. 3.22; [21]). In our examples *r* will be calculated as follows:

$$r = (Ti/Zr)_A/(Ti/Zr)_B$$

and

$$r = \left(^{144}Nd/^{86}Sr\right)_A/\left(^{144}Nd/^{86}Sr\right)_B$$

Similar to what we observed when calculating the mixture value of a single isotope, the ratios of a mixture will depend only on the relative fractions of the two sources, if the denominator ratios are the same or if the denominator is the same (as is very often the case with Pearce element ratios).

All these calculations are very useful and are applied in a variety of ways and for a variety of problematics.

For instance, we can deduce the concentrations and proportions of the sources participating in a mixture if we have a range of mixed samples with high compositional variability. We can generate mixing equations for specific purposes, such as the commonly used element versus ratio mixture calculations, by adapting the general equation developed by Langmuir et al. [21]. Finally, these calculations are used in a variety of situations, not only in alteration quantification, but also in magma mixing or mantle reservoir mixing.

3.5 Geochronology

In the most general terms, geochronology is defined as the art and science of discovering the age of an event that affected a rock. As such, we will consider geochronology in this chapter, even though it is performed on selected minerals and very rarely on whole rock.

3.5.1 General Principles of Geochronology

Earlier, in ▶ Chap. 1, we discussed that radioactive isotopes disintegrate at a constant and known speed, meaning that if we know speed and the amounts of parent and daughter isotopes in a sample, we can calculate its age by using the general geochronology equation. For example, we can rewrite this equation for the ^{235}U–^{207}Pb parent–daughter couple (☐ Table 3.2):

$$Age_{^{207}Pb/^{235}U} = \frac{1}{\lambda_{^{235}U}}\left(^{207}Pb/_{^{235}U} + 1\right)$$

We can also write a very similar equation for the other uranium isotope, ^{208}U, which disintegrates to ^{206}Pb, giving us two ages for the uranium–lead series or for any other parent–daughter couple (☐ Table 3.2).

The general principle of absolute (or isotope) geochronology is fairly straightforward. In practice, though, the application of the method and the interpretations involved can be complex due to several complicating factors. For the method to work, several conditions must be satisfied, the most important of which is that the mineral

☐ **Table 3.2** Parent–daughter couples most commonly used in absolute (isotope) geochronology. Data from Brownlow [6]

Parent isotope	Daughter isotope	Main disintegration path	Parent isotope abundance (%)	Half-life (years)	Disintegration constant λ (year⁻¹)	Minerals typically incorporating the parent isotope
^{40}K	^{40}Ar	$\beta^-, \varepsilon\chi$	0.001167	1.19×10^{10}	5.5492×10^{-10}	Any K-rich mineral
^{87}Rb	^{87}Sr	β^-	27.8346	48.8×10^9	1.42×10^{-11}	Any K-rich mineral
^{147}Sm	^{143}Nd	α	15.0	1.06×10^{11}	6.54×10^{-12}	Various
^{176}Lu	^{176}Hf	β^-	2.6	3.6×10^{10}	1.867×10^{-11}	Garnet, phosphates
^{235}U	^{207}Pb	α	99.2743	7.07×10^8	9.8571×10^{-10}	Zircon, monazite, apatite
^{238}U	^{206}Pb	α	0.7200	4.47×10^9	1.55125×10^{-10}	Zircon, monazite, apatite
^{232}Th	^{208}Pb	α	100.0	1.4×10^{10}	4.948×10^{-11}	Zircon, monazite, apatite

system analysed remains a closed system, i.e., that there is no loss or gain of the parent or daughter isotopes.

3.5.2 Radiogenic Product Retention and Closure Temperature

A major interpretation complication could arise from the way the daughter isotope is accumulated into the mineral analysed. Whereas the parent isotope is most commonly chemically bound within the crystalline structure of the mineral, this is not the case for the daughter isotope, which is often mechanically retained but without being chemically bound. This is particularly true with the ^{40}Ar, which is produced by the disintegration of ^{40}K and is a noble gas. As a result, the daughter isotope will tend to leave the mineral system, most commonly by diffusion. The main factor, among several, controlling the diffusion of a chemical element in a crystalline structure is temperature: the higher the temperature, the higher the diffusion rate and the faster the daughter element will leave. If, for example, a mineral formed at a high temperature (e.g., crystalized in a magma chamber), the ambient temperature may have been so high that that most of the daughter isotope left the system by diffusion. As the rock containing the mineral cooled down, the diffusion rate became lower and lower, to the point where most—and then all—of the daughter isotope was retained and started accumulating within the mineral analysed. This is a critical point for the understanding of isotope geochronology: when we analyse a mineral and calculate an absolute age for it, that age corresponds to the time when the mineral's isotopic system became closed (because it cooled sufficiently enough to retain all of the daughter isotope) and not to the time when the mineral formed (◘ Fig. 3.23). Indeed, there are many cases when the mineral formed several millions of years before the system became closed and the daughter isotopes started to cumulate.

The temperature below which the entirety of the daughter isotope is retained is called closure temperature (or blocking temperature) and is different for each mineral and for each isotopic system (◘ Table 3.3). This means that if we date a rock using two different minerals and two different isotopic systems (e.g., zircon for U–Pb and muscovite for K–Ar), we will likely find two different ages. These ages may be very different, and one of the two may be close to the formation of the mineral. Therefore we must be very careful when we interpret the data.

There are two specific implications that arise from the concept of closure temperature. Firstly, if a mineral was formed at temperatures close to or below the closure temperatures for the particular isotopic system used for dating, then the age obtained will be that of the mineral formation. For instance, if a zircon is formed at 850 °C, then its age obtained by U–Pb dating will be identical to the crystallization age for that zircon, as its closure temperature for Pb is higher. As another example, if a muscovite is formed at 300 °C, the K–Ar age obtained will be the crystallization age for this mineral, as the Ar closure temperature for muscovite is approximately 350 °C. Thus, we can have access to the actual formation of a mineral and a rock, on condition that we select the appropriate isotopic system and the appropriate mineral.

On the other hand, a mineral that was heated at some point during its history—during a metamorphic episode,

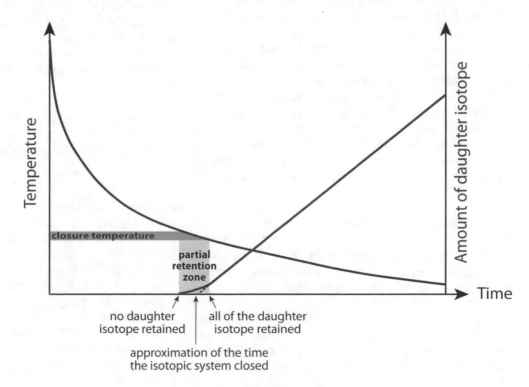

◘ **Fig. 3.23** Graphic representation of the notion of closure temperature and the corresponding isotopic age. The isotopic system will become closed only below a certain temperature and is different for each isotope and each mineral. The time that elapsed before the system became closed cannot be found on the basis of a single geochronological method

◘ Table 3.3 Typical closure temperatures for different minerals and different parent–daughter systems

System	Mineral	Approximate closure temperature (°C)
^{40}K-^{40}Ar	Hornblende	550
	Muscovite	350
	Biotite	300
	Feldspar	200
^{87}Rb-^{87}Sr	Muscovite	450
	Biotite	350
^{235}U-^{207}Pb and ^{238}U-^{206}Pb	Zircon	900
	Monazite	800
	Titanite	600
	Apatite	450
	Rutile	400

The exact and specific closure temperatures can be sometimes difficult to assess, as they are fairly variable and depend on many factors, including grain size and chemical composition of the mineral, the cooling rate, and possibly pressure

for instance—at temperatures above the closure temperatures for the specific isotopic system, would lose some or all of the radiogenic daughter isotope, partially or completely resetting the isotopic "clock" of the mineral. If all of the radiogenic daughter isotope had been lost, then the age we would obtain would be that for the subsequent closure of the system, when the rock cooled again below the closure temperature; in other words we would obtain the age of the reheating event. Thus, the interpretation of geochronological data must always integrate the known evolution of the rocks studied.

The major consequence of the closure temperature concept is that we must always consider what the event is that we are dating: initial formation, cooling below a certain temperature, or reheating at some point after initial system closure. We must remember that we do not obtain the age of a rock or a mineral, but rather the age of an event affecting that rock or mineral; that event resulted in the closure (or the reopening and second closure) of the isotopic system of the mineral analysed. This distinction is an important one, and we should always keep it in mind.

3.5.3 Correction for Common Lead

Another significant complication that arises when applying this method is the fact that the mineral could incorporate, during its formation, some amount of the radiogenic daughter isotope. In some cases, we have good grounds to assume that no daughter isotope was initially present (e.g., when the mineral could not incorporate it into its crystalline structure), but often we suspect that at least some small amount is present. In other words, the daughter isotope that we measure may be not only the product of radioactive disintegration of the parent isotope, but also originally present in the mineral, as is often the case with lead:

$$^{207}Pb_{measured\ now} = ^{207}Pb_{radiogenic} + ^{207}Pb_{original}$$

There are two methods to calculate and subtract the original amount of daughter isotope: the common Pb correction and the isochron method.

The common lead correction relies on the existence of ^{204}Pb, or common lead, which is a lead isotope that is stable and non-radiogenic: its amount on Earth is constant and is the same as it was when the Earth initially formed (this is why we also call it *primordial*). Crucially, some parts of the radiogenic ^{207}Pb and ^{206}Pb were also initially present, but their amounts grew over time as more was produced by the disintegration of ^{235}U and ^{238}U, respectively. At the moment when the mineral (for instance, zircon) formed, it incorporated some amounts of ^{204}Pb, ^{207}Pb, and ^{235}U (we will consider only one of the two U series, but it will work in exactly the same way in both cases). Most importantly, the ratio ^{207}Pb/^{204}Pb for that time is known, as radiogenic lead/common lead ratios are well known for the entire history of the Earth. By using the ^{207}Pb/^{204}Pb ratio for the mineral formation time, we can account for the amount of ^{207}Pb that was incorporated into the mineral at that time:

$$^{207}Pb_{radiogenic} = ^{207}Pb_{measured\ now} - ^{207}Pb_{original} = ^{207}Pb_{measured\ now} - R_t{}^{204}Pb$$

where $R_t = ^{207}$Pb/^{204}Pb at the time the mineral formed.

The main difficulty here is that we do not initially know the formation time of the mineral, so we must use an iterative approach: we start by making the best possible guess of the age, calculate an age and then recalculate a second or a third age, each time using the age from the previous iteration.

The *isochron* method also relies on the existence of ^{204}Pb. We will use the general age equation reformulated for lead:

$$^{207}Pb_{now} = ^{207}Pb_{original} + ^{235}U_{now}(e^{\lambda t} - 1)$$

and then divide each isotope into it by ^{204}Pb:

$$\left(^{207}Pb/^{204}Pb\right)_{now} = \left(^{207}Pb/^{204}Pb\right)_{original} + \left(^{235}U/^{204}Pb\right)_{now}(e^{\lambda t} - 1)$$

In this new equation, only the $\left(^{207}Pb/^{204}Pb\right)_{original}$ is not known, whereas the other two ratios are measured. We will also notice that this is a linear equation $y = a + bx$, where.

$y = \left(^{207}Pb/^{204}Pb\right)_{now}$, which is measured.

$x = \left(^{235}U/^{204}Pb\right)_{now}$, which is also measured.

$a = \left(^{207}Pb/^{204}Pb\right)_{original}$, which is the intercept of the line at $x = 0$, and

$b = (e^{\lambda t} - 1)$, which is the slope of the line (◘ Fig. 3.23).

This method relies on the analysis of several minerals that were formed at the same time, but incorporated dif-

Fig. 3.24 U–Pb isochron diagram. Several different minerals (or whole rock samples) with different initial U–Pb ratios are used. As the ^{235}U progressively disintegrates to ^{207}Pb, the ^{235}U/^{204}Pb ratio will decrease and the ^{207}P/^{204}Pb ratio will increase. This results in a higher slope of the isochron line, corresponding to a higher age. The initial ^{207}P/^{204}Pb ratio remains the same, of course, and can be derived from the intercept of the isochron at ^{235}U/^{204}Pb = 0. The minerals' isotopic composition will always be on the same line

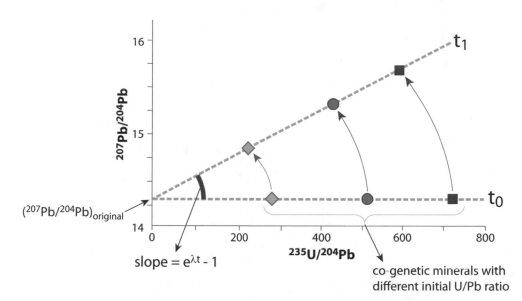

ferent amounts of uranium. Once the slope of the *isochron* ("line of the same age") is obtained, we can calculate the age of the rock which the minerals analysed are part of.

This is a very powerful and reliable method and is very commonly used, not only for the U–Pb systems, but also for the ^{87}Rb-^{87}Sr, ^{147}Sm-^{143}Nd, and ^{176}Lu-^{176}Hf ones and several others. We can rewrite the isochron equation for ^{176}Lu-^{176}Hf system, for instance, by using the stable and non-radiogenic ^{177}Hf (**Fig. 3.24**):

$$\left(^{176}\text{Hf}/^{177}\text{Hf}\right)_{now} = \left(^{176}\text{Hf}/^{177}\text{Hf}\right)_{original} + \left(^{176}\text{Lu}/^{177}\text{Hf}\right)_{now}(e^{\lambda t} - 1)$$

3.5.4 Concordia Diagram

Let us now return to the U–Pb system, which has two parent–daughter pairs, ^{235}U-^{207}Pb and ^{238}U-^{206}Pb. Once the common Pb has been subtracted, the two age equations can be written:

$$\text{Age}_{207Pb/235U} = \frac{1}{\lambda_{235U}}\left(^{207}Pb^*/_{235}U + 1\right)$$

and

$$\text{Age}_{206Pb/238U} = \frac{1}{\lambda_{238U}}\left(^{206}Pb^*/_{238}U + 1\right)$$

where * denotes lead generated only by the disintegration of *U* or radiogenic lead.

The fact that two ages can be derived from the U–Pb system is unique and gives us a powerful tool for deciphering the history of a sample. This tool is called the Concordia diagram (**Fig. 3.22**). The evolution of the terrestrial ^{207}Pb/^{235}U and ^{206}Pb/^{238}U ratios is known, as they are based on the disintegration constants given in **Table 3.2**. If the two U–Pb ages obtained on the same mineral, typically a zircon, are the same, then this is the age when the isotopic system had initially closed (typically at the formation of the mineral), *or* the age when

the system was closed after being completely reset during a reheating event. However, the Pb closure temperature for zircon is fairly high (**Table 3.3**), meaning that we rarely reach temperatures that are able to completely reset zircon's U–Pb system. However, zircon is a very resistant mineral and a single crystal can often participate in numerous igneous or metamorphic events, or both, so that its isotopic system will be affected by partial Pb* loss. In that case, the obtained ^{206}Pb*/^{238}U and ^{207}Pb*/^{235}U ratios will not lie on the Concordia line, but on a straight line, called Discordia, that connects two Concordia/Discordia intercepts (**Fig. 3.25**). The most common interpretation of such a situation is that the upper intercept represents the initial formation age of the mineral and the lower intercept, the age of the perturbation (thermal) event causing the partial Pb* loss.

Finally, we can consider another age calculation issuing from the U–Pb series: the ^{207}Pb/^{206}Pb age. If we divide the two U–Pb age equations by each other and take into account that the ^{238}U/^{235}U ratio is known and constant at 137.88, we come up with the ^{207}Pb/^{206}Pb age equation:

$$\frac{^{207}\text{Pb}^*}{^{206}\text{Pb}^*} = \frac{1}{137.88}\frac{e^{\lambda_{235}t} - 1}{e^{\lambda_{238}t} - 1}$$

In other words, we can analyse only Pb isotopes in a particular sample and still be able to obtain an age derived from the U–Pb disintegration series and interpreted in the same way as the U–Pb ages. This specific method requires the subtraction of common lead. However there is another option to account for it, and that is the ^{207}Pb/^{206}Pb isochron. If we take the two U–Pb isochron equations and divide them by each other, we obtain an equation involving only Pb isotopes:

$$\frac{\left(^{207}Pb/_{204}Pb\right)_{now} - \left(^{207}Pb/_{204}Pb\right)_{original}}{\left(^{206}Pb/_{204}Pb\right)_{now} - \left(^{206}Pb/_{204}Pb\right)_{original}} = \frac{1}{137.88}\frac{e^{\lambda_{235}t} - 1}{e^{\lambda_{238}t} - 1}$$

3

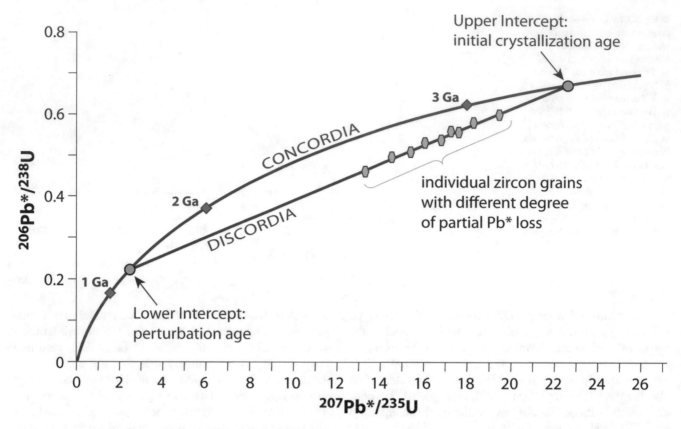

□ **Fig. 3.25**   Concordia diagram, using both $^{206}Pb*/^{238}U$ and $^{207}Pb*/^{235}U$ ratios. The Concordia line is defined as the line on which the two U–Pb ages are the same, whereas the Discordia line is situated between two ages, namely the initial crystallizations age, at the upper intercept, and the perturbation age leading to partial Pb* loss, at the lower intercept

In that manner, if we plot our analyses in a diagram of $^{207}Pb/^{204}Pb$ versus $^{206}Pb/^{204}Pb$, we will obtain an isochron whose slope corresponds to the age of the sample.

To conclude, the U–Pb disintegration series are very powerful and versatile, as they can be used in different ways leading to different ages and can also inform us about the timing of specific events that affected the sample analysed.

The combined U–Pb–K–Ar approach can best be illustrated by the following example. The McLean granitic pluton from the western part of the Grenville orogeny (Ontario, Canada) was first dated by the U–Pb method applied on zircon grains, which yielded an upper intercept age of 1070 ± 7 Ma; the $^{207}Pb/^{206}Pb$ age calculated using the same analyses was 1069 ± 2 Ma ([8], Fig. 3.26). Amphibole, biotite, and potassic feldspar were also extracted and dated by the $^{40}Ar/^{39}Ar$ method (a method derived from the K–Ar method), and the ages obtained were approximately 1060 Ma, 990 Ma, and 790 Ma, respectively (author's data). Closure temperatures were estimated at approximately 500 °C for amphibole, 300 °C for biotite, and 200 °C for K-feldspar. Using these data, it was possible to reconstruct the thermal evolution of the McLean pluton (□ Fig. 3.26). The initial emplacement of the pluton was approximated to the zircon upper intercept age at 1070 Ma. This was followed by rapid cooling, from ca. 800 to ca. 500 °C in

about 10 Ma, probably because the host rock into which the pluton was emplaced was considerably cooler. From there on, the cooling of the pluton slowed down significantly: it lost 100 °C from biotite to K-feldspar, a span of approximately 200 Ma. This is a classic example of thermochronology, or the science of studying the thermal evolution of rocks. It has many applications, such as orogenic and tectonic studies. Most importantly, it illustrates the principle that any isotopic age corresponds to an event, in most cases the time when the isotopic system was closed, or in other words the time when the rock cooled below the specific closure temperature(s).

3.6   Summary

Lithogeochemistry, or the study of whole rock chemical composition, has many practical applications. We can use major, minor, or trace elements or isotopes, for a variety of purposes. Firstly, we can use the total alkali vs. silica (TAS) diagram, to classify any igneous rock, and the Rare Earth Elements (REE) patterns to estimate their degree of differentiation. Using major or trace elements in the appropriate diagrams, we can derive the tectonic setting in which rocks are formed. We can also study the different Earth and mantle reservoirs, using a variety of radiogenic isotopes.

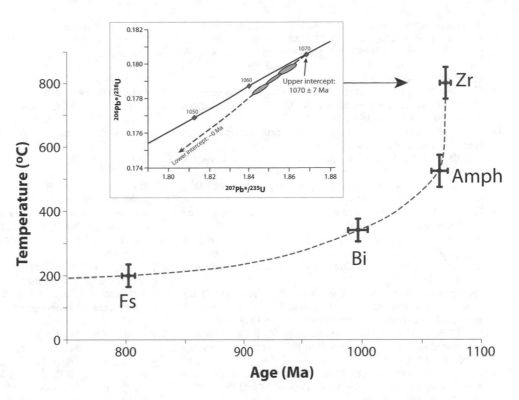

■ **Fig. 3.26** Emplacement age and thermal evolution of the McLean granitic pluton (west Grenville orogeny, Canada), derived from U–Pb dating of zircons (inset) and $^{40}Ar/^{39}Ar$ dating of amphibole (Amph), biotite (Bi), and potassic feldspar (Fs). After a quick initial cooling, the subsequent cooling was much slower, corresponding to the thermal history of the hosting tectonic block. This example illustrates vividly that different geochronological methods applied on different minerals will give different ages, which have to be interpreted very carefully

Whole rock geochemistry is particularly valuable for the quantification of hydrothermal alteration, very useful in mineral exploration. For that purpose we can use isocon diagrams of the Pearce element ratios.

Finally, absolute, or isotope, geochronology relies on the known rate of disintegration of a particular isotope and provides with the age of specific events that affected a rock. Although simple in its general principle, geochronology is subject of several interpretation complications and requires meticulous analysis and careful interpretation.

❓ Exercises

Q3.1 Using the data provided in the Q3.1 MS Excel file, construct the REE spectra for the Earth major reservoirs. Firstly, normalize the concentrations by those of the provided C1 chondrite and then plot the normalized values for all REEs. The Y-axis should be in logarithmic scale. Comment on the shape of the REE spectra, and interpret them in terms of igneous processes.

Q3.2 Using the data provided in the Q3.2 MS Excel file, (1) calculate the U–Pb ratios and construct a Concordia diagram (using increments of 50 Ma), and (2) plot the Discordia line and estimate the ages of the upper and the lower intercepts.

Q3.3 Use the whole rock data provided in the Q3.3 MS Excel file to geochemically define the 5 igneous rock samples listed. Use the appropriate diagrams given in this chapter, and comment on the rocks' type, name, main Earth reservoir, main igneous rock series, and their likely geodynamic setting.

Q3.4 Using the same data (the Q3.3 MS Excel file), construct isocon diagrams comparing the geochemistry of (a) samples 1 and 5 and (b) samples 2 and 3. Outline the chemical differences between the samples that are compared with each other, and comment on the likely causes for these differences.

References

1. Albarede F (2009) Geochemistry, an introduction, 2nd edn. Cambridge University Press
2. Bao Zh, Zhao Zh, Guha J, Williams-Jones AE (2004) HFSE, REE, and PGE geochemistry of three sedimentary rock-hosted disseminated gold deposits in southwestern Guizhou Province, China. Geochem J 38:363–381
3. Barker F (1979) Trondhjemite: definition, environment and hypotheses of origin. In: Barker F (ed) Trondhjemites, dacites and related rocks. Elsevier, Amsterdam, pp 1–12
4. Bhatia MR (1983) Plate tectonics and geochemical composition of sandstones. J Geol 91:611–627
5. Bhatia MR, Crook KAW (1986) Trace element characteristics of graywackes and tectonic setting discrimination of sedimenatry basins. Contrib Miner Petrol 92:181–193
6. Brownlow AH (1979) Geochemistry. Prentice-Hall, 498 pp
7. Cox KG, Bell JD, Pankhurst RJ (1979) The interpretation of igneous rocks. George, Allen and Unwin, London
8. Davidson A, van Breemen O (2000) Late Grenvillian granite plutons in the Central Metasedimentary Belt, Grenville Province, southeastern Ontario. Geological Survey of Canada, Current Research 2000-F5, radiogenic age and isotopic studies. Report 13, 9 pp
9. De la Roche H, Leterrier J, Grande Claude P, Marshall M (1980) A classification of volcanic and plutonic rocks using R1–R2

diagrams and major elements analyses—its relationships and current nomenclature. Chem Geol 29:183–210

10. Dosso L, Murthy VR (1980) A Nd isotopes study of the Kerguelen Islands: inferences on enriched oceanic mantle sources. Earth Planet Sci Lett 48:268–276

11. Elderfield H, Greaves MJ (1981) Negative cerium anomalies in the rare earth element patterns of oceanic ferromanganese nodules. Earth Planet Sci Lett 55:163–170

12. Elderfield H, Greaves MJ (1982) The rare earth element elements in seawater. Nature 296:214–219

13. Estrade G, Salvi S, Béziat D, Rakotovao S, Rakotondrazafy R (2014) REE and HFSE mineralization in peralkaline granites of the Ambohimirahavavy alkaline complex, Ampasindava peninsula, Madagascar. J Afr Earth Sc 94:141–155

14. Faure G (1998) Principles and applications of Geochemistry, 2nd edn. Prentice Hall, New York

15. Gao S, Luo M-C, Zhan, B-R, Zhang H-F, Ha Y-W, Zhao Z-D, Hu Y-K (1998) Chemical composition of the continental crust as revealed by studies in East China. Geochimica et Cosmochimica Acta 62, 1959–1975

16. Harmond RS, Hoefs J (1995) Oxygen isotope heterogeneity of the mantle deduced from global 18O systematics of basalts from different geotectonic settings. Contrib Miner Petrol 120:95–114

17. Hauri EH, Shimizu N, Dieu JJ, Hart SR (1993) Evidence for hotspot-related carbonatite metasomatism in the oceanic upper mantle. Nature 365:221–227

18. Herron MM (1988) Geochemical classification of terrigenous sands and shales from core log data. J Sediment Petrol 58:820–829

19. Hofmann A, White WM (1982) Mantle plumes from ancient oceanic crust. Earth Planet Sci Lett 57:421–436

20. Kemp AIS, Hawkesworth CJ (2004) Granitic perspectives on the generation and secular evolution of the continental crust. In: Holland HD, Turekian KK (eds) Treatise on geochemistry, vol 3. Elsevier, Amsterdam, pp 349–410

21. Langmuir CH, Vocke RD, Hanson GN, Hart SR (1978) A general mixing equation with applications to Icelandic basalts. Earth Planet Sci Lett 37:380–392

22. Le Maitre RW, Bateman P, Dudek A, KellerJ LL, Bas MJ, Sabine PA, Schmidt RC, Sorensen H, Streckeisen A, Wolley AR, Zanettin B (1989) A classification of igneous rocks and glossary of term: Recommendations of the International Union of Geological Sciences Subcommission on the Systematics of Igneous Rocks. Blackwell Scientific Publications, Oxford

23. McQueen KG, Whitbread MA (2009) Lithogeochemical vectors to ore: a study of the Elura Zn-Pb-Ag deposit, Cobar, NSW. In: 24th International Applied Geochemistry Symposium, Fredricton, New Brunswick, Canada, May 2009

24. Middleton GV (1960) Chemical composition of sandstones. Bull Geol Soc Am 71:109–126

25. Mullen ED (1983) MnO/TiO$_2$/P$_2$O$_5$: a minor element discriminant for basaltic rocks of oceanic environment and its implications for petrogenesis. Eartehr Planet Sci Lett 62:53–62

26. Pecerillo R, Taylor SR (1976) Geochemistry of eocene calc-alkaline volcanic rocks from the Kastamonu area, northern Turkey. Contrib Miner Petrol 58:63–81

27. Pettijohn FJ, Potter PE, Siever R (1972) Sand and sandstones. Springer, New York

28. Pearce JA, Norry ML (1979) Petrogenetic implications of Ti, Zr, Y and Nb variations in volcanic rocks. Contrib Mineral Petrol 69:33–47

29. Pearce JA (1982) Trace element characteristics of lavas from destructive plate margins. In: Thorpe RS (ed) Andesites: orogenic andesites and related rocks. Wiley, pp 525–548

30. Pearce JA (1996) A user's guide to basalt discrimination diagrams. In: Wyman DA (eds) Trace element geochemistry of volcanic

rocks; applications for massive sulphide exploration. Geological Association of Canada, Short Course Notes, vol 12, pp 79–113

31. Pearce JA, Cann JR (1973) Tectonic setting of basic volcanic rocks determined using trace element analysis. Earth Planet Sci Lett 19:290–300

32. Pearce TH (1968) A contribution to the theory of variation diagrams. Contrib Miner Petrol 19:142–157

33. Pearce J, Nigel Harris N, Tindle AG (1984) Trace Element Discrimination Diagrams for the Tectonic Interpretation of Granitic Rocks. J Petrol 25:956–983

34. Potter PE (1978) Petrology and chemistry of modern big river sands. J Geol 86:423–449

35. Roser BP, Korsch RJ (1986) Provenance signatures of sandstone-mudstone suites determined using discriminant function analysis of major-element data. Chem Geol 67:119–139

36. Rudnick RR, Fountain DM (1995) Nature and composition of the continental crust: a lower crustal perspective. Rev Geophys 33:267–309

37. Stanley C (2017) Molar element ratio analysis of lithogeochemical data: a toolbox for use in mineral exploration and mining. In: Tschirhart V, Thomas MD (2017) Proceedings of Exploration 17: Sixth Decennial International Conference on Mineral Exploration, pp 471–494 (2017)

38. Stanley CR, Madeisky HE (1996) Lithogeochemical exploration for metasomatic zones associated with hydrothermal mineral deposits using molar element ratio analysis. In: Introduction, Lithogeochemical Exploration Research Project, Mineral Deposit Research Unit, University of British Columbia, Short Course Notes, 200 p

39. Su N, Yang S, Guo Y, Yue W, Wang X, Yin P, Huang X (2017) Revisit of rare earth element fractionation during chemical weathering and river sediment transport. Geochem Geophys Geosyst 18:935–955

40. Taylor HP Jr (1978) Oxygen and hydrogen isotope studies of plutonic granitic rocks. Earth Planet Sci Lett 38:177–210

41. Taylor SR, McLennan SM (1985) The continental crust: its composition and evolution. Blackwell Scientific Publications, Oxford

42. Turekian KK, Wedepohl KH (1961) Distribution of the elements in some major units of the Earth's crust. Geol Soc Am Bull 72:175–192

43. Vinogradov AP (1962) Average contents of chemical elements in the principal types of igneous rocks of the Earth's crust. Geochemistry 1962:641–664

44. Walter MJ (2004) Melt Extraction and Compositional Variability in Mantle Lithosphere. In: Holland HD, Turrekian KK (eds) Treatise on geochemistry, vol 2. Elsevier, Amsterdam, pp 363–394

45. Wilson M (1989) Igneous petrogenesis. Unwin Hyman, London

46. White WM (1985) Sources of oceanic basalts: radiogenic isotope evidence. Geology 13:115–118

47. White WM (2013) Geochemistry. Wiley-Blackwell, Oxford

48. Zindler A, Hart SR (1986) Chemical geodynamics. Annu Rev Earth Planet Sci 14:493 571

Further Reading

49. **Geochemistry, an introduction**, by Albarede F (2009) Chapter 11 The solid Earth. Cambridge University Press. ISBN 987-0-521-88079-4.

50. **Geochemistry**, by W.M. White (2013) Chapter 11 The geochemistry of the Solid Earth. Wiley-Blackwell, ISBN 987-0-470-65667-9.

51. **Isotope Geochemistry**, by White WM (2015) Wiley-Blackwell. ISBN 987-0-470-65670-9.

52. **Using Geochemical Data: Evaluation, Presentation, Interpretation**, by Rollinson HR (1993) Chapters 3 and 4. Pearson Education, ISBN 0-582-06701-4.

Geochemical Exploration

Contents

4.1 Introduction and General Principles – 62
4.1.1 Definition and Purpose of Geochemical Exploration – 62
4.1.2 The Concept of Source-Transport-Trap – 62
4.1.3 Anomaly – 63

4.2 Distribution and Dispersion of Elements and Formation of Surface Anomalies – 63
4.2.1 Geochemical Behaviour of the Elements – 63
4.2.2 Pathfinder Elements – 63
4.2.3 The Mobilization of Elements – 63
4.2.4 Formation of a Surface Anomaly – 66

4.3 Understanding the Formation and Evolution of a Mineral Deposit – 68
4.3.1 Fluids Involved in Deposit Formation – 68
4.3.2 Physical Conditions of Deposit Formation – 69
4.3.3 Chemical Conditions of Deposit Formation – 70
4.3.4 Processes Involved in Deposit Formation – 71
4.3.5 Geochronology in Metallogeny – 72

4.4 Detection of Surface Anomalies – 72
4.4.1 Quantification of Alteration – 72
4.4.2 Soil Sampling – 73
4.4.3 Sampling of Stream Sediments – 76
4.4.4 Sampling of Vegetation – 76
4.4.5 Sampling of Surface Waters – 76
4.4.6 Survey Design – 76

4.5 Isotopes in Exploration – 77

4.6 Data Interpretation: Spatial Geostatistics – 78
4.6.1 Summary Statistics – 78
4.6.2 Spatial Data Visualization – 79
4.6.3 Kriging – 81

4.7 Summary – 82

References – 82

Electronic supplementary material The online version of this chapter (▶ https://doi.org/10.1007/978-3-030-72453-5_4) contains supplementary material, which is available to authorized users.

© Springer Nature Switzerland AG 2021
P. Alexandre, *Practical Geochemistry*,
Springer Textbooks in Earth Sciences, Geography and Environment,
https://doi.org/10.1007/978-3-030-72453-5_4

4

4.1 Introduction and General Principles

4.1.1 Definition and Purpose of Geochemical Exploration

Each mineral deposit type has a set of characteristics—often clearly defined and well understood—some of which will be very different from the host rocks and will be therefore used in mineral exploration. The deposits' physical characteristics will be used in geophysical exploration, their lithological, mineralogical, and structural characteristics will be used in prospecting, and their chemical characteristics will be used in geochemical exploration. Accordingly, in the most general terms, geochemical exploration can be defined as *the science of using geochemical knowledge and tools to discover mineral deposits that are under cover*. In that sense, geochemical exploration is a *remote sensing tool*—not unlike geophysics—relying on the geochemical characteristics and behaviour of the elements.

Another, and not insignificant, purpose of geochemical exploration is *to understand how deposits form* and *under what conditions*, for which geochemistry is a very powerful tool. This knowledge will be directly used, in turn, to improve exploration activities; this is an indirect way of finding hidden mineral deposits.

Mineral deposits represent physical objects—ore bodies—with extreme enrichment in a specific element relative to the average continental crust values; the enrichment factors typically vary between 2 and 5 orders of magnitude. Therefore, it is very useful to consider mineral deposits as, first and foremost, *extreme geochemical anomalies*. It is also useful to remember that the commodity of interest is a specific chemical element (e.g., Cu), or sometimes a combination of two or three (e.g., Pb–Zn–Au), or rarely more, in the so-called polymetallic deposits. Therefore, when we study min-

eral deposits or explore for them, we must always consider the geochemical characteristics and behaviour of the commodity of interest, described in ▶ Chap. 1. It is of extreme importance that we thoroughly understand the characteristics and behaviour of the element of interest—our commodity—to be able to explore efficiently. Without this knowledge and understanding, we are bound to make some very silly mistakes and never find a deposit. Therefore, the "first law of geochemical exploration" should be *Know Thy Commodity*. Let us take a simple but perfectly real example: Li is a mobile highly incompatible lithophile element (▢ Figs. 1.4, 1.9 and 1.10). Thus, we will much more likely find it in an extremely differentiated lithology (e.g., pegmatite) that has been subject to hydrothermal activity than in an ultramafic rock that is very fresh. In other words, the simple knowledge of a chemical element's geochemical behaviour can be a very powerful and *direct* exploration tool.

4.1.2 The Concept of Source-Transport-Trap

Another interesting and important concept in geochemical exploration is that of *source-transport-trap*. It is a common and appropriate conceptual framework for the study of mineral deposits, in which we consider the initial source of the commodity of interest, its mobilization (physical, by erosion, or chemical, by dissolution), its transport (as a solid, liquid, or gas), and the focusing of the flow to a suitable trap (physical or chemical, or both), where the element is immobilized to form a deposit. However, the exact same idea can be applied to mineral exploration (▢ Fig. 4.1). Some—typically small—amount of the commodity of interest is mobilized from a deposit under cover, transported, and then trapped in a suitable *surface trap* to make a surface anomaly that we will attempt to discover as a direct

▢ **Fig. 4.1** The concept of source-transport-trap can be used when studying the formation of mineral deposits, but also when studying the formation of surface anomalies. The processes in the two parts are exactly the same. Both aspects are parts of geochemical exploration: the first is understanding how deposits form and under what conditions, and the second is finding surface anomalies as direct guides to hidden deposits

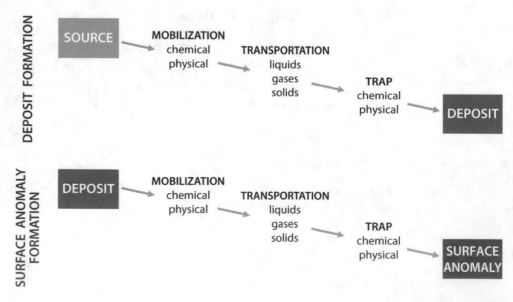

guide to the deposit. This concept, when applied to exploration geochemistry, indicates that we must consider how a chemical element is mobilized, transported, and trapped (or immobilized) in the surface environment: considerations that all directly relate to the geochemical characteristics and behaviour of the elements, hence the importance of knowing them well. In this chapter, we will consider the different stages of the formation of a surface anomaly and how they are affected by a variety of factors, as well how to efficiently detect a surface anomaly.

4.1.3 Anomaly

This leads us to the idea of anomaly. In geochemical exploration we can use the following general definition: *"An anomaly is a chemical concentration that is significantly different from the local background"*. In ▶ Chap. 2 we proposed a mathematical formulation for anomaly (◨ Fig. 2.16), but there may be other formulations and ideas; ultimately the important parts are *"significantly* different" and *"local* background" (◨ Fig. 4.2). This connects to what we mentioned earlier: deposits are extreme geochemical anomalies, and their surface expressions are also geochemical anomalies relative to the local background.

4.2 Distribution and Dispersion of Elements and Formation of Surface Anomalies

4.2.1 Geochemical Behaviour of the Elements

The distribution of chemical elements in igneous rocks, and to a lesser degree in metamorphic rocks, is mostly controlled by their geochemical characteristics and behaviour, specifically their compatibility and their mobility in the presence of water. Generally speaking, the compatibility of an element results in a clear and direct relationship between the degree of differentiation of an igneous rock and the element's concentration. As

mentioned earlier, in the example of Li, we can predict—with a fair degree of certainty—how enriched in a particular element a rock will be with a specific degree of differentiation. On the other hand, the amount of water will also affect the mobility of elements, albeit the relationship between element mobility and concentration is a bit more complicated. Nevertheless, these two characteristics of the chemical elements, compatibility and mobility, often result in specific rocks being more likely, or "favourable", to contain a certain deposit type or certain commodity. Let us consider two examples, that of volcanogenic massive sulphides (VMS) and of porphyry copper deposits. Experience has demonstrated that for VMS deposits, the most favourable rock types are sub-alkaline volcanic rocks with an intermediate degree of differentiation, such as andesite and dacite, and possibly rhyodacite. As for porphyry copper deposits, the most favourable rocks are (shallow level) granitoids from ocean-continent subduction: the increased volatiles content of the subducting oceanic plate not only causes partial melting, but also mobilises copper and thus generates copper-rich magmas. It is important to know, as much as possible, the typical—or favourable—rock types and geodynamic setting for the commodity and for the deposit type that we are exploring for, as this will be of immense help early in the (greenfield) exploration program.

4.2.2 Pathfinder Elements

From the point of view of geochemical behaviour of the elements stems the notion of associated or *"pathfinder"* elements: elements which, while not necessarily economically significant, appear in association with the main commodity in a specific deposit type. The significance of these elements is that they can be used during geochemical exploration, as they may lead to a surface anomaly and to a deposit in a way very similar to that of the main commodity. The pathfinder elements are often defined empirically and can vary to a certain degree within the same deposit type. ◨ Table 4.1 offers a selection of pathfinder elements for a few deposit types.

4.2.3 The Mobilization of Elements

The mobilization of elements from the deposit (at depth) to the surface, where they will form a geochemical anomaly, occurs principally in three ways: mechanical (e.g., erosion), chemical (e.g., weathering), and biogenic (e.g., by vegetation). Let us consider these here, starting with mechanical dispersion.

Mechanical dispersion
There are several ways mechanical dispersion works (◨ Fig. 4.3). The major ways are surface erosion

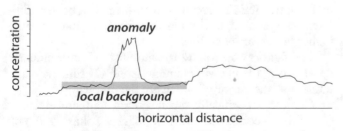

◨ **Fig. 4.2** General definition of geochemical anomaly as concentrations significantly different from the local background. The distal, or regional, background is mostly irrelevant

4

□ **Table 4.1** Examples of element associations for selected deposit types. It is noticeable that the elements that occur together share the same chemical characteristics and geochemical behaviour (e.g., compatible elements), as well as belonging to the same geochemical reservoir (e.g., chalcophile elements)

Deposit type	Main elements	Associated elements
Magmatic deposits		
Chromite ores (e.g., Bushveld)	Cr	Ni, Fe, Mg
Layered magnetite (e.g., Bushveld)	Fe	V, Ti, P
Immiscible Cu and Ni sulphides	Cu, Ni, S	Pt, Co, As, Au
Pt, Ni, and Cu in layered intrusion	Pt, Ni, Cu	Cr, Co, S
Immiscible Fe and Ti oxides	Fe, Ti	P
Nb–Ta carbonatite	Nb, Ta	Na, Zr, P
Rare-metal pegmatite	Be, Li, Cs, Rb	B, U, Th, REE
Sedimentary deposits		
Copper shale (Kupferschiefer)	Cu, S	Ag, Zn, Pb, Co, Ni, Cd
Copper sandstone	Cu, S	Ag, Co, Ni
Hydrothermal deposits		
Porphyry copper (e.g., Bingham)	Cu, S	Mo, Au, Ag, Re, As, Zn, K
Porphyry molybdenum (e.g., Climax)	Mo, S	W, Sn, F, Cu
Skarn—magnetite	Fe	Cu, Co, S
Skarn—Cu (e.g., Yerington)	Cu, Fe, S	Au, Ag
Skarn—Pb–Zn (Hanover)	Pb, Zn, S	Cu, Co
Skarn—W–Mo–Sn (e.g., Bishop)	W, Mo, Sn	F, S, Cu, Be, Bi
Base metal veins	Pb, Zn, Cu, S	Ag, Au, As, Sb, Mn
Sn–W greisens	Sn, W	Cu, Mo, Bi, Li, Rb, Si, Re, F, B
Sn sulphide veins	Sn, S	Cu, Pb, Zn, Ag, Sb
Co–Ni–As vein	Co, Ni, Ag, S	As, Sb, Bi, U
Epithermal precious metals	Au, Ag	Sb, As, Hg, Te, Se, S, U
Uranium vein	U	Mo, Pb, F
VMS—Cu	Cu, S	Zn, Au
VMS—Zn–Cu–Pb	Zn, Pb, Cu, S	Ag, Ba, Au, As
Au-As-rich iron formation	Au, As, S	Sb
Mississippi Valley Pb–Zn	Pb, Zn, S	Ba, F, Cd, Cu, Ni, Co
Sandstone U	U	Se, Mo, V, Cu, Pb
Red-bed Cu	Cu, S	Ag, Pb

dispersion, glacial and wind dispersion, and weathering dispersion (and, in a more limited way, bioturbation-mediated dispersion). Three important issues should be considered here. The first is the fact that that during mechanical dispersion the commodity of interest is mobilized—by erosion or weathering—most likely as a part of a mineral phase. It is transported as such by surface or subsurface movement or eroded material, and is immobilized by deposition in a low-energy depositional environment. Thus, it is the characteristics of the mineral phase containing the commodity of interest that will be the controlling factor in the formation of the surface anomaly, and not the characteristics of the element of interest, or at least not directly. Secondly, additional complications may—and very often do—arise when a more complex surface depositional environment is present. For instance, a barren glacial deposit or colluvium can develop on top of the residual overburden in weathered profile, thus hiding the anomaly produced by the weathering-mediated mechanical dispersion. Finally, the surface mechanical dispersion-produced anomaly is often displaced, at least a short distance (but sometimes as far as several kilometres), relative to the ore body. This will be true for anomalies produced by other modes of dispersion and underlines the fact that a surface anomaly is virtually never found directly above a deposit at depth.

Chemical dispersion

A second major dispersion mechanism is chemical dispersion, where the element of interest are placed in an aqueous solution and transported as dissolved species. In this case, the element's own geochemical behaviour, and specifically its mobility in the presence of water (▶ Chap. 1, □ Fig. 1.10), directly conditions its dispersion and subsequent immobilization in the surface environment. With chemical dispersion, we are very interested in the movement of water through soil and through rocks. When an element disperses using ground water in overburden, two cases are present (□ Fig. 4.4): vertical and lateral movement of water. When an element is dispersed through a rock, water moves either by diffusion or by mass transport utilizing pre-existing fluid flow channels such as fractures. It is important to notice that diffusion is typically one order of magnitude slower than mass transport. The surface anomalies produced by these two fluid transport mechanisms will be very different, larger and low-grade for diffusion, narrow and higher grade for mass flow.

The factors controlling the mobility of the elements, apart from their own characteristics (□ Fig. 1.9) are numerous, the major of which are the following:

- Climate—specifically mean annual temperature and rainfall—combined with topography has a direct relationship with the type and degree of weathering and dispersion. High temperatures with large volumes of water are the optimum conditions for

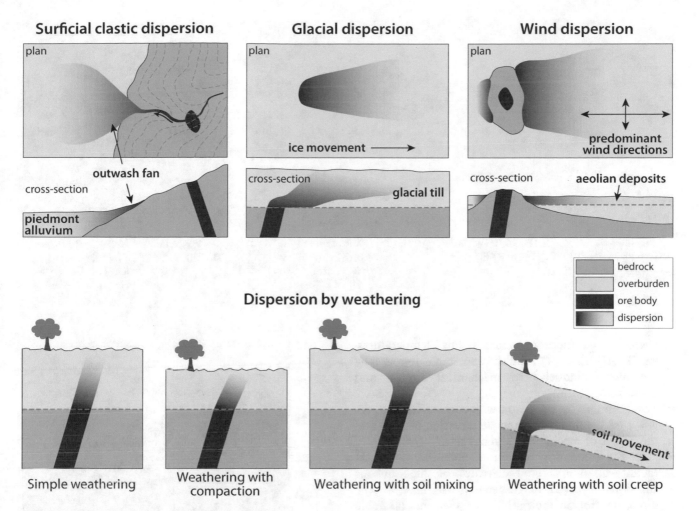

Surficial clastic dispersion

plan

cross-section

outwash fan

piedmont alluvium

Glacial dispersion

plan

ice movement →

cross-section

glacial till

Wind dispersion

plan

predominant wind directions

cross-section

aeolian deposits

bedrock
overburden
ore body
dispersion

Dispersion by weathering

Simple weathering

Weathering with compaction

Weathering with soil mixing

Weathering with soil creep

soil movement

◻ **Fig. 4.3** Typical examples of mechanical dispersion mechanisms of a commodity, or element of interest, in the surficial environment. Crucially, the commodity is mobilized, transported, and immobilized as a part of a mineral phase

◻ **Fig. 4.4** Vertical and lateral water movement in soil and the two different chemical dispersion patterns produced

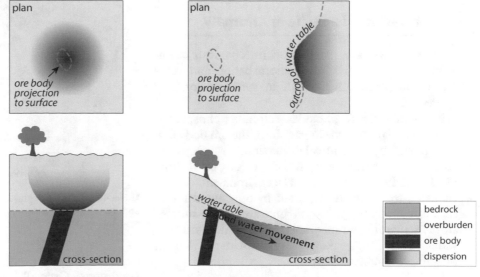

plan

ore body projection to surface

plan

ore body projection to surface

outcrop of water table

water table

ground water movement

Upward movement of moisture

Lateral ground water flow

cross-section

cross-section

bedrock
overburden
ore body
dispersion

◘ Table 4.2   Mobility of elements as function of Eh and pH. Most cations tend to be more immobile in reducing alkaline conditions

Relative mobility	Conditions			
	Oxidizing	Reducing	Acid	Neutral to alkaline
Very high	B, Br, Cl, I, S	B, Cl, I	B, Br, Cl, I, S	B, Br, Cl, I, S, Mo, V, U, Se, Re
High	Mo, V, U, Se, Re, Ca, Na, Mg, F, Sr, Ra, Zn	Ca, Na, Mg, F, Sr, Ra	Mo, V, U, Se, Re, Ca, Na, Mg, F, Sr, Ra, Zn, Cu, Co, Ni, Hg, Ag, Au	Ca, Na, Mg, F, Sr, Ra
Medium	Cu, Co, Ni, Hg, Ag, Au, As, Cd		As, Cd	As, Cd
Low	Si, P, K, Pb, Li, Rb, Ba, Be, Bi, Sb, Ge, Cs, Tl	Si, P, K, Fe, Mn	Si, P, K, Pb, Li, Rb, Ba, Be, Bi, Sb, Ge, Cs, Tl, Fe, Mn	Si, P, K, Pb, Li, Rb, Ba, Be, Bi, Sb, Ge, Cs, Tl, Fe, Mn
Very low to immobile	Fe, Mn, Al, Ti, Sn, Te, W, Nb, Ta, Pt, Cr, Zr, Th, REE	Al, Ti, Sn, Te, W, Nb, Ta, Pt, Cr, Zr, Th, REE, S, B, Mo, V, U, Se, Re, Zn, Cu, Co, Ni, Hg, Ag, Au, As, Cd, Pb, Li, Rb, Ba, Be, Bi, Sb, Ge, Cs, Tl	Al, Ti, Sn, Te, W, Nb, Ta, Pt, Cr, Zr, Th, REE	Al, Ti, Sn, Te, W, Nb, Ta, Pt, Cr, Zr, Th, REE, Zn, Cu, Co, Ni, Hg, Ag, Au

chemical weathering, whereas low temperatures, small amounts of rainfall and high topography are more conducive to mechanical erosion and dispersion.

— Vegetation, in coordination with climate, has a direct effect on the formation and preservation of soils and the degree of mechanical and chemical dispersion.

— The Eh (oxidation–reduction potential) and pH (acidity–alkalinity) characteristics of the aqueous solutions involved in dispersion will have a very direct control on the mobility of elements (◘ Table 4.2). Waters from different environments will have distinct Eh and pH (◘ Fig. 4.5) and thus will tend to mobilize different groups of elements.

4.2.4 Formation of a Surface Anomaly

The two main ways of dispersion (mechanical, chemical), in addition to the minor bioturbation and vegetation remobilization, will result in surface anomalies composed of:

— primary minerals (e.g., oxides, sulphides, phosphates, native metals) sourced directly from the ore body and transported by mechanical dispersion;

— secondary minerals (e.g., Mn-Fe oxides or hydroxides, salts) formed in the surface environment;

— elements that were transported by aqueous solution (or more rarely by gas) and then trapped in a suitable surface environment medium (soil, organic matter);

— water bodies (bogs, lakes, streams) where the element of interest remains in solution.

The important consideration here is that the same ore body can—and often does—produce a variety of surface expressions (surface anomalies) as function of the dispersion mode and the predominant environmental

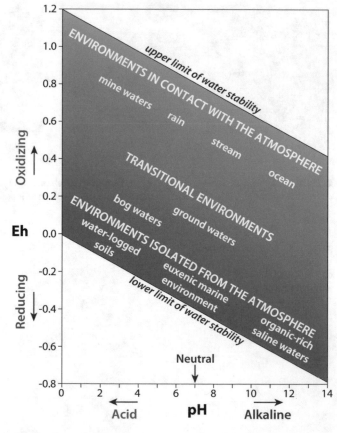

◘ Fig. 4.5   Eh–pH diagram showing the full range of water stability and the position of waters from specific reservoirs. After Baas-Becking et al. [5], modified

conditions. Crucially, the different surface anomalies will not be at the same place and will not have the same intensity, underlining the paramount importance of both the thorough planning of an exploration campaign and the careful interpretation of the data obtained.

The surface anomalies produced by mechanical dispersion—by erosion—will tend to form in low-energy settings, such as the bottom of streams or lakes, in alluvium, etc., where the mobilized particles will accumulate. On the other hand, the surface anomalies produced by chemical dispersion—by mobilization in an aqueous solution—will form when the dissolved elements (as cations or oxyanions) become trapped in a suitable surface medium, most likely inorganic matter (e.g., clay minerals in soils) or organic matter (e.g., vegetation such a trees or bushes).

Adsorption

The trapping of elements on the surface of minerals in the soil is called *adsorption*; it can be chemical (chemisorption, when a chemical bond is formed) or physical (physisorption, when ions are retained by the weak van der Waals force). Other processes can also result in the trapping of elements in surface media (e.g., ion exchange, precipitation, hydrolysis, polymerization), and together with adsorption, we use the general term *sorption*. The root cause of adsorption is the bond deficiency at the surface of minerals (in itself an unstable site), compensated through the—weak and temporary—bonding with ions present in the adjacent aqueous solution. By far the most favourable minerals to trap dissolved elements on their surface are those with the largest bond deficiency (or "structural charge"), such as clay minerals, Fe and Mn oxides, and carbonates, all of which are present, to a certain degree, in the soil. From this point of view, soil is the best and most popular surface sampling medium, in the absence of outcropping lithology, which is the most common situation in exploration.

The degree to which an ion will be adsorbed is conditioned by a number of environmental factors, such as weather (e.g., mean annual temperature and rainfall), orography, and the pH of the aqueous solution. Regarding the pH, there is a very strong dependence on how much of the dissolved ions are adsorbed on the acidity or alkalinity of the solution (■ Fig. 4.6). The main reason for this is that the H^+ ions available in the solution will be in direct competition with other cations for the surface bonding sites of the minerals present; and the OH^- molecules will be in direct competition with the dissolved anions.

Another very important factor affecting adsorption is the surface charge. As an example, anions will tend to adsorb to Fe oxides even at low pH due to their positively charged surface, whereas cations will prefer to adsorb to negatively charged surfaces even at neutral to alkaline pH conditions. Given that most surfaces (e.g., oxides, clays, microbial surfaces) will have some charge, surface charge is an important factor that affects the adsorption, in conjunction with the pH, leading to wide variations in the adsorption versus pH curves (■ Fig. 4.6).

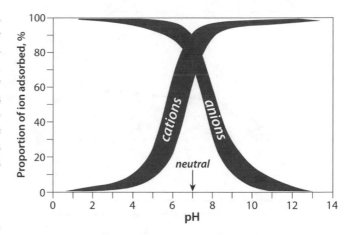

■ **Fig. 4.6** General form of the dependence of the amount of adsorbed ions on the pH of the aqueous solution: cations will adsorb at neutral and alkaline pH (less competition from H^+), whereas anions will adsorb at neutral and acid pH (less competition from OH^-), assuming no variations in charge of the adsorbing surfaces. The specific shape of the relationships will vary to a fair extent and will be unique for each specific combination dissolved ion—adsorbing mineral surface and its charge

Thus, it can be generalized that the lower the pH (more H^+ present in the solution), the lower the amount of adsorbed cations will be; and the higher the pH (more OH^- present in the solution), the lower the amount of adsorbed anions will be. The exact relationships between amounts adsorbed as function of pH are variable and unique for each combination of dissolved ion (and its speciation) and adsorbing mineral, and specifically its surface charge. Crucially, the largest variations in amount adsorbed are near neutral pH (■ Fig. 4.6), which is the most common one in natural surface waters. In other words, even minute changes in the pH of the solution (groundwater in our case) can lead to significant changes in the proportion of ions adsorbed. Looking at ■ Fig. 4.6, we realize that, as an example, a change in pH of 0.3 (6.7–7) can lead to an increase of adsorbed cation of approximately 15%. Thus, it is imperative that when soil samples are collected, the pH of the water present (even if there isn't much of it) is measured, which is simple, fast, and inexpensive. The results, in terms of presence or absence of a geochemical soil anomaly, may be due uniquely to local variation in pH and thus lead to the detection of "false positives" or "false negatives". If the pH at the sampling site is known, any geochemical variation can be verified by checking the pH variability and confirmed as real or rejected as due to pH variation.

Biochemical anomalies

As mentioned earlier, mobile ions can be trapped in vegetation, typically trees and bushes, and thus form biogeochemical anomalies. In plants, various ions serve as macro-nutrients (e.g., K, P), or micronutrients (e.g., Fe, base metals). The idea is that the concentration of an

4

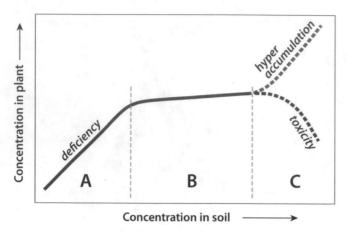

● **Fig. 4.7** Generalized relations between the concentration of an element in soil and its concentration in a plant. Along segment A, the concentration in the plant reflects that of the underlying soil; this is the ideal situation for exploration. Along segment B, an exclusion or storage mechanism is operating and there is no further bioaccumulation, in spite of increased concentrations in the soil. With increasing concentrations in the soil, along segment C, hyper-accumulation or toxicity may develop

element in the plant will represent more or less directly the concentrations of the element in the soil. While this relationship is generally true, it is subject to variations due to a variety of factors, such as weather (including small-scale variations), the health of the plant, its age, etc., but also as a function of the concentration of the element of interest in the soil.

The main advantage of this geochemical exploration method is that the broad root systems of a large tree can provide extensive and representative samples, in particular considering that the depth of roots puts the samples closer to the possible source. It is also a relatively easy and inexpensive method, as the samples are easily collected, transported, and analysed.

The use of plants in exploration has been mentioned more than a century ago and has been extensively used since the 1950s, mostly in Canada and Russia, but also in any place with little rock exposure and sufficient plant cover. However, interpretation of the results is not always straightforward, as the factors controlling bioaccumulation are numerous. Ideally, plants of strictly the same species, of similar age, and in good health are collected to minimize biological variations; little—except for careful data interpretation—can be done when the plant does not faithfully represent the element concentration in the soil (along segments B or C; ● Fig. 4.7).

4.3 Understanding the Formation and Evolution of a Mineral Deposit

While not strictly and directly related to mineral exploration, understanding how a deposit—or a deposit type—formed and evolved is always very necessary

for effective exploration. There is a great variability of deposits and deposit types for any commodity, and the explorationist has to have a very clear understanding of the object that is being explored for in order to employ the most appropriate and effective exploration techniques, be it prospecting, geophysics, or geochemistry, or a judicious combination of these. This understanding includes the deposit or the deposit type's specific lithological and mineralogical characteristics, its physical and chemical characteristics, its size and morphology, and the geochemical characteristics and the geochemical behaviour of the commodity—the chemical element—of interest. Both aspects (deposit characteristics and element's geochemical behaviour) are generally relatively well known, and the most effective exploration efforts will add the other necessary piece of information: the knowledge of the deposit formation.

Knowing how a deposit formed and evolved involves gaining a specific and clear understanding of the processes and conditions (physical and chemical) during the mobilization, transport, and immobilization (trapping) of the commodity to form a deposit (● Fig. 4.1), as well as the timing of the relevant events. Once a clear model for the formation of a deposit or a deposit type has been formulated, the ultimate step—and the one that will have the most direct impact on exploration efficiency—is to define the critical factors that control the formation of a deposit: this is the understanding that can most directly be translated into effective exploration tools.

Whereas geophysics is a superb and powerful tool for mineral exploration, it is geochemistry—in combination with petrology and mineralogy—that is the most useful set of methods leading to a clear understanding of the formation of a deposit. Geochemistry can help with all aspects of deposit formation, and we will consider these in turn. Let us first consider the example of stable isotopes and the fluids that participated in the formation of a deposit, both by producing the appropriate host rock alteration and by transporting the commodity of interest to the deposition site.

4.3.1 Fluids Involved in Deposit Formation

When we speak of fluids, we much more likely think of aqueous fluids, and these have two main components—chemically speaking—hydrogen and oxygen. Through a series of processes (predominantly evaporation and precipitation) occurring under varying conditions in the hydrosphere and specifically during the global water cycle, each water found in Earth's main reservoirs will have its own clear and specific hydrogen and oxygen isotopic compositions ([2], ● Fig. 4.8). Most saline waters (oceans and seas) will be found in a single spot (at δ^2H and $\delta^{18}O = 0‰$). All fresh waters

Fig. 4.8 Oxygen and hydrogen isotopic compositions for the main water reservoirs on Earth (ocean water, meteoric—or precipitation—water, and their evaporated versions), for waters contained in metamorphic and magmatic rocks, and for meteoric waters that interacted with metamorphic or magmatic rocks. Note that all fresh water (lakes, streams, groundwater) on the continents will have the isotopic composition of the meteoric water that produced it by precipitation. After Alexandre [2], modified

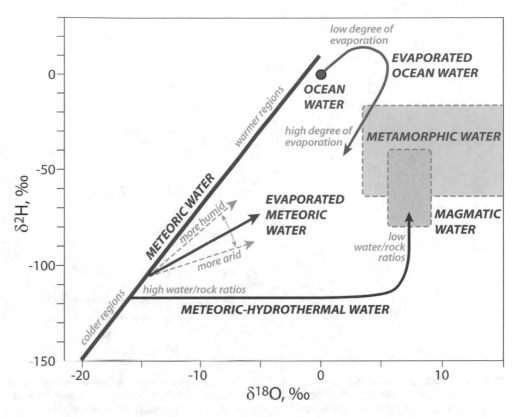

(precipitation, lakes and streams, groundwater) are placed on a single line (the Global Meteoric Water Line). The waters contained in metamorphic and magmatic waters have their own fields, and meteoric water that interacted with them has a specific composition corresponding to the isotopic mixture between the initial meteoric water and the specific rocks it interacted with (**Fig. 4.8**). Evaporated ocean waters and river or lake waters have also their specific isotopic compositions (**Fig. 4.8**).

In other words, if we measure the hydrogen and oxygen isotope compositions of gangue and ore minerals from a mineral deposit, and calculate the isotopic compositions of the waters from which the minerals precipitated, we can have a very clear idea of the source, or origin, of these waters (**Fig. 4.9**), but also have some idea of the processes involved in the deposit formation. For instance, the alterations observed at the Butte porphyry copper deposit in Montana, USA, were produced uniquely by meteoric waters, with a decreasing water/rock ratio as alteration processes advanced from argillic to sericite to potassic alteration (**Fig. 4.9**). On the other hand, the Mississippi Valley-type (MVP) deposits in Illinois, USA, were formed uniquely by evaporated low-latitude meteoric water (**Fig. 4.9**). There are many other examples of hydrogen and oxygen isotope compositions revealing the source of the waters that participated in the formation of a deposit: indeed, this is the most powerful and unambiguous way to discover the fluid history of a deposit or a deposit type.

4.3.2 Physical Conditions of Deposit Formation

Now that we know what the fluids that were involved in the formation of our deposit were, we can turn our attention to the specific physical conditions (pressure and temperature) of deposit formation. There are two major ways of discovering these conditions: mineral chemistry and stable isotopes. The first method exploits the dependence of element substitutions in minerals on pressure and temperature, as we discussed in ► Chap. 1 (► Sect. 1.6). There, we noted that the extent of substitution is affected by the physical and chemical conditions of the system, with higher temperatures allowing for higher substitution tolerance and higher pressure resulting in lower tolerance for substitutions. Based on this idea and on empirical observations, several geothermometers and geobarometers have been developed and are particularly useful in finding the temperature at which a specific mineral formed. Specific geothermometers and geobarometers exist for a variety of minerals, such as illite and chlorite (e.g., [7–9]), quartz (Ti-in-Q, [19, 21]), sphalerite [13], and many others.

Another way to find the precipitation temperature of a mineral is to use the fractionation of stable isotopes, which is a function of temperature. If two minerals formed at the same time—as confirmed by microscopic observations—and at isotopic equilibrium (e.g., quartz and calcite for vein gold deposits, or pyrite-pyrrhotite for a MVT deposit), we can derive their formation temperature based on the fractionation equation for the two specific minerals (**Fig. 4.10**).

4

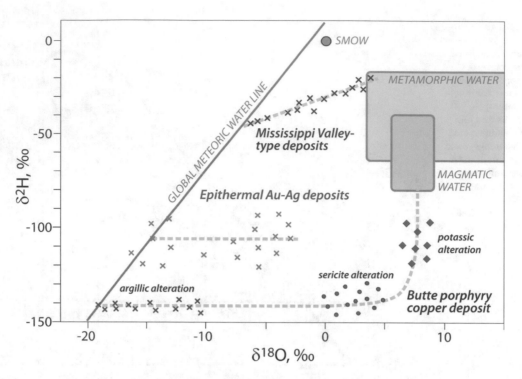

☐ **Fig. 4.9** Isotopic composition of ore-forming fluids for three different types of deposits from the USA: the epithermal (low temperature) Au–Ag deposits; the Mississippi Valley-type deposits in the Illinois; and the Butte porphyry copper deposit in Montana. Each type of deposit was formed by a different water: the Mississippi Valley-type deposits by the strongly evaporated meteoric water in the Illinois Basin, the epithermal Au–Ag deposits by high-latitude meteoric water that had undergone weak interaction with the host rocks (high water/rock ratio), and the Butte deposit by strongly evolved meteoric water from even higher latitudes. In this last case, the progression of alteration, from low-temperature argillic to the high-temperature potassic alteration, can be followed by the progressively lower water/rock ratio and eventual isotopic equilibration with the host rock, the Butte quartz monzonite. Data from Kyser [15]

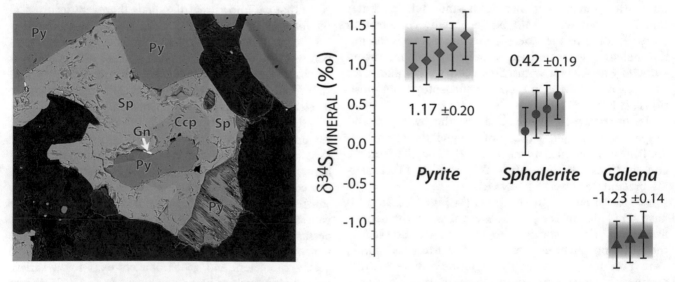

☐ **Fig. 4.10** Example of sulphur isotope measurements applied to the study of ore deposits. The samples come from a base metals and gold volcanogenic massive sulphide deposit in northern Manitoba, Canada (after Alexandre et al. [4], modified). When the $\Delta^{34}S$ between the three possible couples of minerals, pyrite-sphalerite (Py–Sp), pyrite-galena (Py–Gn), and sphalerite-galena (Sp–Gn), are considered, a temperature of about 400 °C can be derived, based on the fractionation factors between the minerals

4.3.3 **Chemical Conditions of Deposit Formation**

The chemical composition of gangue minerals can also provide clues about the chemistry of the

fluids from which the minerals precipitated (as alluded to in ▶ Chap. 1, ▶ Sect. 1.6 substitutions). The chemical composition of some minerals (e.g., carbonates, clays) faithfully represents that of the fluid they formed from, assuming all elements are saturated in it. We can also use the fluid inclusions present in some minerals (e.g., quartz, calcite, sphalerite, fluorite, halite, apatite), as they are a direct representation of the fluids from which these minerals precipitated. As such, fluid inclusions can provide direct indications of the pressure and temperature conditions of mineral formation. In practical terms, the sample (a "thick" section of approximately 200 μm across) is super-cooled until all components in the fluid inclusions are frozen, and then slowly warmed up to room temperature. The phase transitions observed, such as first melting and last melting, provide direct indications of the nature of the dissolved chloride species (e.g., NaCl, KCL, MgCl$_2$, CaCl$_2$, etc.) and their concentration (or the salinity of the fluid), respectively. Then, the sample is heated to relatively high temperatures (usually 300 °C) and the temperature at which all phases homogenize into one phase—typically liquid—is noted: this temperature ("homogenization" temperature) is the minimum temperature of trapping of the fluid. When the salinity information obtained earlier is combined with the minimum temperature estimate, a minimum trapping pressure can also be estimated. Thus, this method—*microthermometry*, as it is called—can provide direct and clear evidence about the salinity of the fluid and the pressure and temperature of formation of the mineral. On the other hand, we can use laser ablation-ICP-MS to extract and directly measure the chemical composition of the fluid, in particular that of its trace elements. These methods, microthermometry and LA-ICP-MS on fluid inclusions, are powerful, albeit complicated analytical tools and are applied predominantly in the research environment by highly qualified specialists; we will not dedicate more time to this topic here, but will recommend a good book for further reading (Fluid Inclusions: Analysis and Interpretation, by Samson et al. [24] (◘ Fig. 4.11).

4.3.4 Processes Involved in Deposit Formation

By now we have a very good idea of the fluids involved in the ore and gangue formation processes (their origin and chemical characteristics) and of the physical conditions (both pressure and temperature) at which these processes occurred. It is now time to turn our attention to the nature of the ore-forming processes, for which we will use the example of uranium deposits. Uranium exists in two major oxidations states, U^{6+} and U^{4+}, and its mobility is highly dependent on the Eh of the system. It is highly mobile—as uranyl complexes—in oxidizing conditions and is highly immobile in reducing conditions (◘ Fig. 1.9), where it precipitates most often as uraninite, UO_2 (most deposits are formed at oxidation-reduction barriers, where uranium is reduced). Uranium is also a highly incompatible element (◘ Fig. 1.10) and is therefore highly enriched in strongly differentiated igneous rocks such as pegmatites, leucogranites, carbonatites, and alaskites. Now, let us contrast uranium with thorium, another highly incompatible element (◘ Fig. 1.10) enriched in the same highly differentiated igneous rocks. However, thorium has only one oxidation state, Th^{4+}, and is largely immobile, including in oxidizing conditions (◘ Fig. 1.9). Therefore, if we compare the concentrations of U and Th in a suite of rocks, we can have a very clear idea of the processes that lead to the formation of a U deposit (◘ Fig. 4.12). If only magmatic processes were involved, then both U and Th will be enriched, and to the same extent (the Th/U ratio will remain the same; ◘ Fig. 4.13). However, if oxidizing fluids were involved to transport uranium and produce a

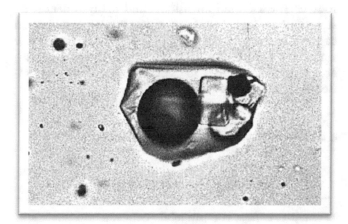

◘ **Fig. 4.11** A typical three-phase liquid–vapour–solid fluid inclusion (approximately 60 μm across). In this case the ore-forming liquid in this inclusion was fairly hot and had such high salinity that several minerals (halite, sylvite, gypsum, and hematite) formed

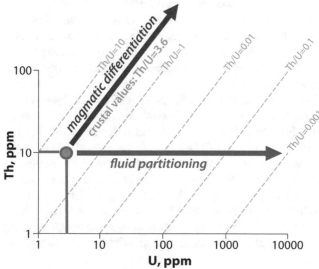

◘ **Fig. 4.12** General relationship between U and Th when enrichment was produced by magmatic differentiation processes (both U and Th are enriched as both are highly incompatible) or through mobilization by oxidizing fluids (Th is immobile, but U is highly mobile in oxidizing fluids as uranyl complexes)

4

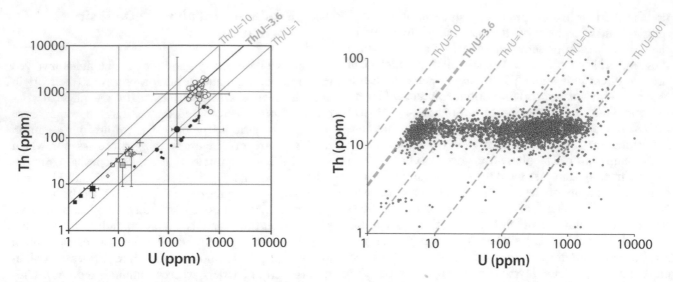

☐ **Fig. 4.13** Relation between Th and U in the different plutonic units of the Ilimaussaq intrusion (Greenland; left) and from the Aricheng Na-metasomatism-related deposit (Guyana; right), clearly demonstrating the different processes leading to economic enrichments: extreme magmatic differentiation for Ilimaussaq and mobilization by oxidizing fluids for Aricheng. Data from Sørensen et al. [18], Bailey [6], Clausen [10], and Alexandre [1]

deposit (as is the case with majority of U deposits), then only U will be enriched as it is the only one in the couple that is mobile under oxidizing conditions (☐ Fig. 4.12).

4.3.5 Geochronology in Metallogeny

Another geochemical method that can provide some ideas about the processes and mechanisms of deposit formation is geochronology. Generally speaking, the goal here is to connect a deposit temporally—and thus causally—to local lithologies: if the deposit has the same age as a local lithological unit (let's say an intrusion), then the deposit might be produced by the processes related to that unit. Let us consider just two examples out of many. The first is that of the giant Chuquicamata Cu and Mo (+Au +Ag) porphyry deposit in northern Chile. Geochronology studies, using the ^{40}Ar/^{39}Ar and Re-Os methods, identified two major hydrothermal events: a high-temperature potassic alteration with chalcopyrite at 33.4 ± 0.3 Ma, followed by lower temperature quartz-sericite alteration with pyrite at 31.1 ± 0.3 Ma. These ages have been directly related to the emplacement ages of the major intrusions of the Chuquicamata Porphyry Complex, dated by the U-Pb method at 36–33 Ma [16]. The interpretation here is that the multiple intrusions provided the thermal energy for the extensive hydrothermal activity that produced the giant deposit. It is indeed the protracted igneous activity here that explains the extent of the hydrothermal activity and the size of the deposit.

Another example is that of the unconformity-related uranium deposits of the Athabasca Basin in Canada. The ore-related clay alterations (illite and chlorite) and the ore itself from different deposits across the basin were dated by the ^{40}Ar/^{39}Ar and U–Pb methods, respec-

tively, and demonstrated that the main ore-formation event at all deposits in the basin occurred at approximately the same time, at about 1.59 Ga [3]. This was interpreted to indicate that it was basin evolution-related processes, and specifically diagenetic processes, that conditioned the formation of the deposits: at the age when the majority of the deposits formed, the basin had reached its maximum depth and therefore its maximum temperature at depth, meaning that the optimum conditions for mobilization of uranium from a variety of detrital phases were reached.

In conclusion, geochronology has the power to provide genetic—or causal—information about mineral deposits, but also to provide suggestions about the relevant processes involved in the formation of the deposits. Combined with the other geochemical methods outlined above (stable isotopes, mineral chemistry, whole rock chemistry, fluid inclusions) and many others (e.g., structural geology, petrology, mineralogy), we are indeed capable of fully understanding when, how, and under what conditions a particular deposit formed and, ideally, able to translate this knowledge into effective exploration tools and techniques.

4.4 Detection of Surface Anomalies

4.4.1 Quantification of Alteration

A major aspect of mineral deposits is their specific alteration assemblage, which can be used as a direct exploration tool. When we have access to rock samples, from rock exposure or through drilling, the whole rock geochemistry can provide clear and

◘ Fig. 4.14 Example of pearce element ratios developed to quantify hydrothermal alterations associated with porphyry Cu–Mo and epithermal Cu–Ag (–Au) deposits in northern Chile. The two main trends are clearly visible, one for magma differentiation (visible as feldspar fractionation) and the other for hydrothermal alteration (samples trending towards and clustering around the muscovite control line, and with further alteration, trending towards the clay control line). After Urqueta et al. [20], modified

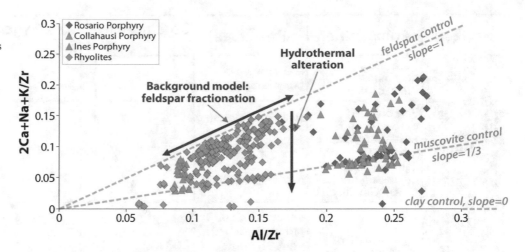

◘ Fig. 4.15 Map of the variation of the alteration factor, defined from Pearce Element Ratios as the slope at which a sample will be found on ◘ Fig. 4.13. The lowest values, associated with strong muscovite alteration, are tightly associated with the four known deposits in the area (numbered) and hint at the possible presence of undiscovered deposits in the south portion of the area. After Urqueta et al. [20], modified

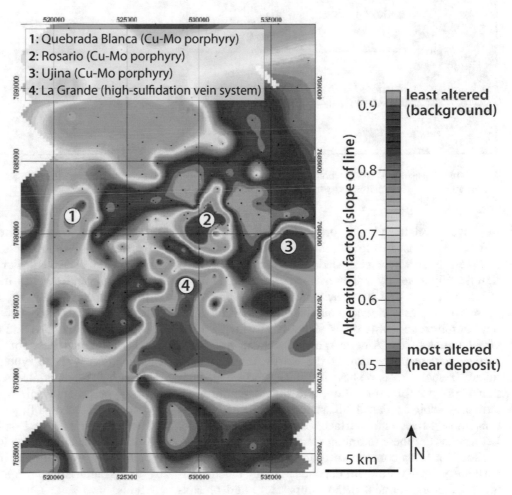

unambiguous information about the type and intensity of that alteration, as discussed in ▶ Chap. 3 (▶ Sect. 3.4; ◘ Fig. 3.18). This information can, in turn, be used as an exploration tool: if we observe the appropriate and favourable type of alteration for the deposit type we are exploring for, and if we observe high intensity of this alteration, then the likelihood of the proximity of a deposit is that much higher. A clear example from exploration for porphyry Cu–Mo and epithermal Cu–Ag (–Au) deposits in northern Chile is shown in ◘ Figs. 4.14

and 4.15, where the intensity of the hydrothermal alteration is quantified as the slope on which any particular sample will lie, with lower slopes indicating higher degree of alteration, and therefore proximity to ore.

4.4.2 Soil Sampling

Lithogeochemistry is a very useful exploration method, but has one major limitation: it is applicable only when

4

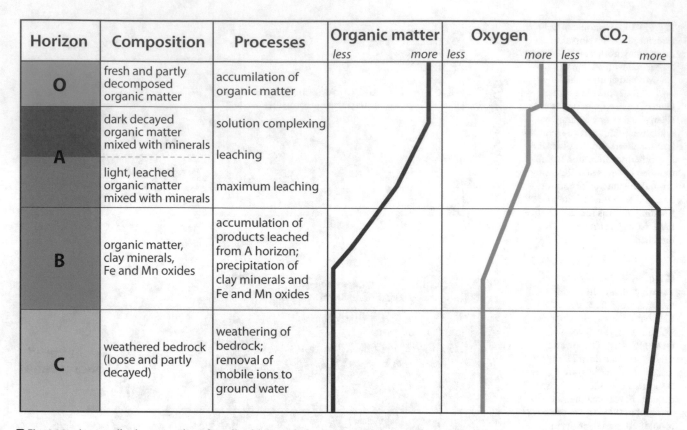

Horizon	Composition	Processes	Organic matter less — more	Oxygen less — more	CO₂ less — more
O	fresh and partly decomposed organic matter	accumilation of organic matter			
A	dark decayed organic matter mixed with minerals	solution complexing			
		leaching			
	light, leached organic matter mixed with minerals	maximum leaching			
B	organic matter, clay minerals, Fe and Mn oxides	accumulation of products leached from A horizon; precipitation of clay minerals and Fe and Mn oxides			
C	weathered bedrock (loose and partly decayed)	weathering of bedrock; removal of mobile ions to ground water			

◼ **Fig. 4.16** A generalized cross section of a soil, with its main horizons, their composition and the main processes occurring, as well as the relative variations of organic matter, oxygen, and CO_2 abundances

we have a relatively high proportion of outcropping bedrock, such as in mountainous regions or in colder climates. Everywhere else, bedrock outcrop represents significantly lower than 10% of the surface and is much lower even than that in many other areas, in particular those where warm climate coincides with low topography. For instance, outcrop is virtually absent from the Great Plains of the American and Canadian prairies, a very significant proportion of the North American continent. In such a case, we have to use any surface medium that is present for exploration, and that is—more often than not—soil or glacial till, stream sediments, vegetation (mostly trees), and surface water (streams or lakes). Let us consider these sampling media in turn.

Soil is a fairly complex medium, with a great deal of variability. Its development depends strongly on environmental factors—annual mean temperature and precipitation—and topography. The composition and thickness of a particular soil are conditioned on these environmental factors and on its age: soil development takes somewhere between 1000 and 10,000 years, or between 5 and 50 mm per 1000 years. Thus, young (or immature) soils in cold mountainous regions with a low amount of precipitation will be very thin, sometimes just a few centimetres, whereas old soils in hot and humid regions may reach as much as 150 m in some extreme cases ("deep" weathering); typical temperate climate soil is somewhere between less than a metre and a few metres thick.

Because of the great variability between soils—and also because of their importance—they are extensively studied and classified. For instance, the US Department of Agriculture has its own classification scheme with 12 "orders" (alfisols, andisols, aridisols, entisols, gelisols, histosols, inceptisols, mollisols, oxisols, spodosols, ultisols, vertisols) and countless subdivisions. The World Reference Base for Soil Resources has another classification with 30 types, while a simpler classification with 9 types is also sometimes used (oxisol, calcisol, gypsisol, gleysol, argillisol, spodosol, vertisol, histisol, and protosol). Ultimately, we are not that interested—or obsessed—with the particular soil classification, even though we have to be fully aware of the type of soil predominant in our exploration area and of its characteristics.

Soils have four main vertical layers, or horizons: the top O horizon, the A and B horizons, and the bottom C horizon (◼ Fig. 4.16). In general terms, O horizon is made predominantly of fresh or weakly decomposed biomass, A horizon is dominated by decomposed organic matter, B horizon is a transitional one, and C horizon is dominated by weathered bedrock.

An important consideration from an exploration point of view is the mineralogical composition of soils. Generally speaking, the most common minerals present in soils are clays (montmorillonite, illite, vermiculite, chlorite, and kaolinite), Fe oxides (goethite, ferrihy-

◘ Table 4.3 Mobility and trapping mechanisms in soil at near neutral pH and high Eh, for a selection of elements. Both the mobility and the trapping mechanisms will vary extensively as function of the Eh and pH of the ground water (see ◘ Table 4.1), as well as environmental factors such as mean annual temperature and precipitation. OM = organicmatter

Element	Oxidation state in soil	Mobility in soil	Trapping mechanism in soil
As	As^{3+} and As^{5+}	Moderately mobile	Chemisorption on Fe and Al oxides at low pH
Ba	Ba^{2+}	Mostly immobile	Precipitation with carbonates, adsorption on clays
Be	Be^{2+}	Mobile	Chemisorption in the presence of OM
Cd	Cd^{2+}	Highly mobile	Precipitation with carbonates
Co	Co^{2+} and Co^{3+}	Moderately immobile	Adsorption on Fe and Mn oxides
Cr	Cr^{3+}	Immobile with organic matter	Substitution in Fe oxides, adsorption on clays
Cu	Cu^{2+}	Moderately immobile	Adsorption on Fe, Mn, and Al oxides
Mn	Mn^{2+}, Mn^{3+}, Mn^{4+}	Immobile	Forms its own insoluble oxides
Mo	Mo^{6+}	Immobile, particularly at low pH	Adsorption on Fe and Al oxides and on OM
Ni	Ni^{2+}	Immobile	OM complexes, precipitation with Fe and Mn oxides
Pb	Pb^{2+}	Immobile, in particular at low Eh	Adsorbs on Mn oxides, binds strongly with OM
Sb	Sb^{3+} and Sb^{5+}	Immobile	Chemisorption on goethite
Se	Se^0, Se^{2-}, Se^{4+}, Se^{6+}	Highly mobile	Immobile as selenides
Ti	Ti^{1+} and Ti^{3+}	Ti^{1+} highly mobile	Ti^{3+} insoluble as Ti_2O_3
V	V^{4+} and V^{3+}	Highly mobile	Adsorption on or substitution in Fe oxides
Zn	Zn^{2+}	Highly mobile	Chemisorbed on oxides

drite, hematite), carbonates (calcite, dolomite), as well as some gypsum and gibbsite. The specific mineralogical composition of any particular soil will be a direct result of the environmental factors mentioned earlier. Importantly, the main minerals in soil (clays, carbonates, and Fe oxides) are also those with the largest bond deficiency (or "structural charge"), meaning that they are the most suitable for surface trapping—or adsorption—of available mobile ions. From this point of view, the soil horizon with the highest amount of minerals will be the most suitable for exploration, and that is typically the B horizon (A horizon is too organic matter-rich, C horizon is too bedrock components-rich). The typical behaviour of some elements in soil is given in ◘ Table 4.3, even though these behaviours will vary enormously as a function of the local Eh and pH and of environmental factors, such as mean annual temperature and precipitation.

In practical terms, soil is collected by first excavating a hole in the ground, ideally exposing all soil horizons including the C horizon. Once the different horizons have been identified, approximately 1–2 kg of sample is taken from the B horizon and placed in a plastic bag. The pH of the sample can be measured—preferably—right away, or later in the laboratory. It is crucial that the sampling personnel is very well trained in soil horizons identification, as mixing soils from different horizons will severely affect the interpretations. At this stage, they should also be extremely careful to avoid contaminating the soil samples: even small personal items, including jewellery, may contribute foreign material into the sample and thus significantly bias the results. Small pieces of rock or organic matter (e.g., bits of wood) are removed by hand right away.

Soil sample preparation consists of drying it and sieving it to a fine fraction, typically under 100 or 50 μm. In some rare cases, the sample is disintegrated—by ultrasound—and the clay fraction (e.g., under 2 microns) is separated for analysis; this is a much more labour-intensive and therefore expensive procedure with some limited benefits relative to avoiding it. Then the sample is analysed, typically in a specialized commercial laboratory. Here, the dissolution technique is of paramount importance and must be a subject of discussion between the exploration company and the analytical laboratory. The fundamental idea is that the element of interest—our commodity—is most often not structurally bound within the crystalline structure of the minerals present in the soil, but is adsorbed on the surfaces or is only weakly bound to it (◘ Table 4.3). Thus, dissolving the whole sample will be counter-productive: the signal coming from the minerals themselves will overwhelm the signal coming from the adsorbed fraction (the one that presumably represents the elements that emanate from a deposit at depth). In other words, in order to maximize the signal from the adsorbed anomalous elements, we should not dissolve the "carrier" minerals, but only mobilize the adsorbed fraction. A series of analytical extraction methods exist for this purpose, and we can use plain distilled water (for the elements that are very weakly adsorbed), or any weak leachant that will not dissolve the adsorbing mineral. This technique is called partial leaching or partial dissolution, and the leachants used include various enzyme leaches (e.g., mobile metal ion, or MMI), NH_4 acetate, NH_4OH, dilute HCl, and any other rather weak acid. On the other end of the spectrum will be the strongest dissolution acids such as *aqua regia* ($3HCl + 1HNO_3$) or the 4-acid combination—HNO_3, HF, $HClO_4$, and HCl—which will achieve near-total dissolution of the sample, something that is not desired in soil exploration. Once the adsorbed or weakly bound elements have been placed in solution, this solution is analysed, most commonly by a quadrupole mass spectrometer; the results are then reported to the exploration company.

4.4.3 Sampling of Stream Sediments

Stream sediments are collected by hand from the bottom of any creek, brook, or river. Stream sediments sampling reflects the mechanical dispersion of minerals from a deposit at depth; the hydrography of the area must be relatively well understood and the order of the sampled tributary must be known. Some sort of sampling device is often used (typically a plastic scoop) and the sample—normally approximately 1 kg—is placed, with no modification, into a plastic bag and sent to the laboratory. There, the sample is subjected to near total dissolution using strong acids, as the element of interest will likely be present in a mineral that was transported from the deposit at depth. The resulting solution is analysed using a quadrupole mass spectrometer; the results are then reported to the exploration company.

4.4.4 Sampling of Vegetation

The next surface medium used in mineral exploration that we will consider is vegetation. Specifically, the most common sampling materials are tree trunk, small branches, and treetop branches. Tree trunks are sampled by hand, using a coring device (◘ Fig. 2.4), which produces a small (typically 7 mm in diameter) core from the tree trunk (the tree is not injured beyond repair and lives on with minimal damage). The sampling is typically done at chest height, to avoid contamination from organic matter in the soil, and for the operator's comfort. The bark on each side is removed, to avoid atmospheric contamination, and the remaining wood is placed in a plastic bag.

When branches are collected, they are taken from a few different places on the tree (or bush), to increase representativity, and placed in a plastic bag. A relatively common method is treetop sampling, often preferred because of the speed of collection using a helicopter, much superior to that of walking on the ground. The very top of the tree is snipped using some sort of clippers, and the sample is placed in a plastic bag.

As with soil samples, extreme care should be taken to avoid any possible contamination, ideally working with laboratory-grade gloves. Another important consideration is that diverse species of trees and trees of varying health intake ions from the soil in different manner and in different amounts. Thus, the operators must identify the most common tree species in the sampling area and use only that species; they also must look out for signs of health and avoid trees that are not in the best health.

The tree samples collected (wood, branches, or treetop) are then sent to the analytical laboratory. There, the wood is most often incinerated and the ashes are analysed by total dissolution and using a quadrupole mass spectrometer. Alternatively, the sample is dissolved using a strong acid and the solution analysed using a quadrupole mass spectrometer; the results are then communicated to the exploration company.

4.4.5 Sampling of Surface Waters

Water samples are by far the easiest to collect (and to analyse). Basically, a water sample is collected into a small plastic bottle, taking care to avoid any contamination, and a few drops of acid are added. The acidification helps prevent adsorption of cations on the walls of the bottle (◘ Fig. 4.6) and to avoid precipitation. The sample is sent to the commercial laboratory where it is analysed, with little preparation or treatment, usually using a quadrupole mass spectrometer.

4.4.6 Survey Design

Let us now consider another important issue, that of survey design. Survey design is a very complex process that takes into account a multitude of factors, some of which pertain to the commodity explored for and the type of deposit it will likely be found in. This includes environmental factors such as the climate and the topography in the sampling area, the size of the area explored, and economic factors such as the time, funds, and personnel available to conduct the geochemical exploration campaign. The purpose of the sampling survey design is to maximize the probability that a sufficient number of good quality samples will be collected to provide a reasonably clear surficial geochemical representation of the area. This will, in turn, allow the detection of any geochemical anomaly that may be present, at an acceptable and economically viable cost. A great many practical issues must be decided beforehand, and some of these are:

— What sampling patterns should be used: regular or irregular grid, or traverse(s), or following geographical features; what is the sampling density?
— What sampling material should be preferentially taken: sampling medium, sample size, initial sampling treatment?

- What is the desired degree of spatial resolution: "first-pass" large scale, or advanced exploration denser sampling program?
- What is the time available to conduct the sampling campaign: fast (a few weeks) or long-term (several months to years)?
- What technology will be used (e.g., helicopters, GPS, hand-held XFR) and what in situ or in-camp analyses will be performed?
- What is the total sampling budget available?
- What are the personnel required?

It is at the early survey design definition stage that the analytical methodology is decided on, in direct relationship with other factors such as the sample type, the budget, the desired timeframe, the desired precision, and so on. At this planning stage, the external commercial laboratory should already be involved in the decision process, as the laboratory personnel are often very knowledgeable about the specific methodology—sample preparation and analysis—best suited for a particular commodity and a particular deposit type, and for the particular types of samples collected.

4.5 Isotopes in Exploration

Stable isotopes can be an actual—direct and effective—exploration tool: both O and S isotopes have been demonstrated to show clear variation as function of proximity to ore, or as function of ore grade. One such case has been documented for the volcanogenic massive sulphide deposits of the Flin Flon-Snow Lake Mineral Belt in northern Manitoba, Canada, where several Zn–Cu–Pb(+Au) deposits have been discovered [17]. Sulphide minerals from several ore bodies and from barren areas were extracted and analysed for sulphur isotopes, and the results show a clear difference between the two ($\blacksquare$ Fig. 4.17). Above a certain $\delta^{34}S$ value, the probability of finding a mineralized ore body decreased sharply. This is possibly explained by the change of sulphur isotopic composition as a function of sulphur's oxidation state: it is consistently higher in oxidized sulphur (as in sulfates) than in reduced sulphur (as in sulphides), by typically about 10 to 20‰. Other examples of stable isotopes applied to exploration include carbon isotopes in oil exploration, where higher $\delta^{13}C$ values of hydrocarbons detected at surface correspond to higher level of hydrocarbon maturation, or oxygen isotopes in gold exploration, where lower $\delta^{18}O$ have been observed to correlate well with the presence of productive gold deposits.

A very similar case was observed in groundwater exploration for base metal deposit in Rajasthan, India. Here, the Hinta prospect, with its significantly lower $\delta^{34}S$ values than the host barren rocks—similar to that in Northern Manitoba ($\blacksquare$ Fig. 4.17)—was clearly reflected in the sampled ground waters ($\blacksquare$ Fig. 4.18). Background ground water $\delta^{34}S$ values vary from 8.5 to 13.5‰ (an average of 10.4‰), whereas the anomalous ground water values range from 6.3 to 7.8‰ (an average 7.0‰). The area where the anomalous ground water values are observed is downhill, in terms of ground water flow, from the deposit ($\blacksquare$ Fig. 4.18). This means that when interpreting water anomalies, we have to be fully aware of the local hydrography in order to correctly interpret

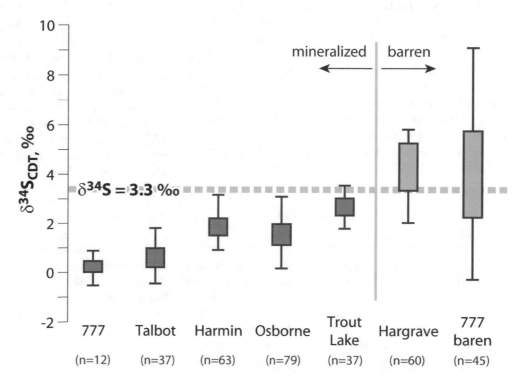

$\blacksquare$ **Fig. 4.17** Sulphur isotopic composition of sulphide minerals from mineralized and barren systems in northern Manitoba, Canada. The chart shows a clear distinction between the two types of systems, possibly due to sulphur's oxidation state, which is itself due to the prevailing Eh conditions during deposit formation [17]

4

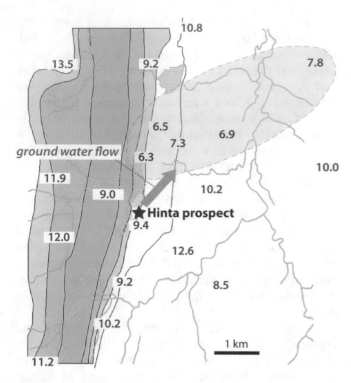

	82	76	104	112	122	21	40	110	115	120
	83	61	110	120	119	75	52	112	116	124
	83	74	96	104	112	92	72	115	117	129
	88	70	105	112	121	64	85	104	113	124
	90	87	95	110	115	108	75	107	117	126
	77	83	86	101	108	114	78	101	120	123
	74	81	85	90	99	101	95	71	128	131
	75	80	82	88	95	98	94	48	138	146
	76	85	73	108	121	143	90	51	136	145
	85	106	46	110	122	109	15	97	134	144

☐ **Fig. 4.18** Sulphur isotopic composition of groundwater from the Hinta Pb–Zn prospect (near Udaipur, Rajasthan, India). The typical local background $\delta^{34}S$ values (varying from 8.5 to 13.5‰) are shown in green; the anomalous $\delta^{34}S$ values resulting from the nearby sulphide mineralization (6.3–7.8‰) are shown in red, clearly demonstrating the effects of the mineralization (red star). Data from de Caritat et al. [12]

☐ **Fig. 4.19** An example of geochemical exploration data, where 100 soil samples were taken on a regular grid of 100 by 100 m and analysed for V. All the obtained data, in terms of V concentrations (in ppm), are reported directly on the map. Data from Isaacs and Srivastva [14], modified

the results and find the source responsible to the anomalous values (i.e., the deposit we are exploring for).

There are plenty of other examples of the direct use of isotopes in exploration, such as the Big Bonanza in the Comstock Lode mining district (Nevada, USA). Here the core of the ore system has significantly higher $\delta^{18}O$ values (0 to 3.8‰) than the surrounding host rocks (–1.0 to –4.4‰; [11]). While this observation is central in deciphering the origin of the deposit as being formed from a large convection cell involving meteoric waters, it also can provide a direct exploration tool: samples with higher $\delta^{18}O$ values are closer to ore. This example, and the other cases outlined earlier, provides clear evidence for the usefulness of isotopes in mineral exploration and not only on the study of mineral deposits.

4.6 Data Interpretation: Spatial Geostatistics

The most common statistical treatment methods were described earlier, in ▶ Chap. 2, ▶ Sect. 2.6. In terms of single variable, the "summary" statistics include the measurement of central tendency and dispersion, and the definition of anomaly. Here, we will devote some time to a very important statistical methodology, namely *spatial geostatistics*, or the study of single variable central

tendency and dispersion as function of space, and how best to visualize data in space—typically in two dimensions—and visualize a surface geochemical anomaly.

4.6.1 Summary Statistics

Let us consider a simple dataset consisting of 100 soil samples taken on a regular grid of 100 by 100 m (☐ Fig. 4.19). A single variable was measured, V, and the concentration results, in ppm, were reported directly on the map and also in a data table. Using these data, the summary statistics can be quickly calculated and communicated in terms of central tendency (mean of 97.7 ppm, median of 101 ppm) and dispersion (minimum of 15 ppm, maximum of 146 ppm, Q_1 of 82 ppm, Q_3 of 116 ppm, P_{10} of 70 ppm, and P_{90} of 128 ppm).

The summary statistics reported above are fairly useful in that they give us a very clear idea of the concentrations of V in the soil in the study area. However, they do not help us at all in finding where the V surface anomaly is situated, which is, after all, the main goal of the exploration endeavours. We can, of course, look at the data as reported in ☐ Fig. 4.19 and attempt to see where the largest V concentrations are situated. This is difficult and unreliable, even in this relatively simple example of 100 data points, and virtually impossible with larger datasets often containing many more than 1000 data points. Therefore, we must find a practical way to reliably visu-

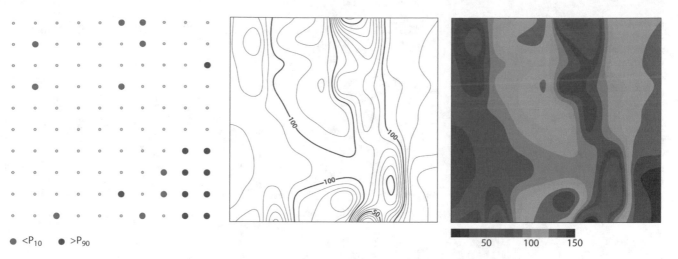

● <P$_{10}$ ● >P$_{90}$

50   100   150

■ **Fig. 4.20** Three examples of the different ways of visualizing the spatial variability of geochemical data, based on the example from ■ Fig. 4.19: (1) colour coding of the highest 10% (>P_{90}) and the lowest 10% (<P_{10}) of the data, (2) lines of equal values, and (3) colour representation of areas of specific categorical concentration ranges

alize the location of the surface anomaly, and this is one of the main purposes of spatial geostatistics.

4.6.2 Spatial Data Visualization

One of the simplest ways to visualize the geochemical data is to use a set of symbols, such as colour coding. For instance, we can use one colour (e.g., blue) for all points below a specific value (e.g., P_{10}, or 70 ppm in our example) and another (red) for those above another value (P_{90}, or 128 ppm; ■ Fig. 4.20). This simple device allows us to clearly see that the largest concentrations are clustered in the SE portion of the study area (■ Fig. 4.20). Instead of colour-coded dots we can use specific numbers or symbols, or we can relate the size of the dot to the concentration (size-coding: small dot for low concentrations, large dot for high concentration), but the basic idea remains the same: the numbers are replaced by a symbol that is, hopefully, easier to detect by the viewer.

Another very common technique to visualize geochemical data is to draw lines of equal concentrations (generally named isopleths or specifically isoconcentration lines), based on the analyses available (■ Fig. 4.20). The same technique can be applied to any numerical values associated with known spatial positions, the most common being the lines of equal elevation used extensively in representing the topography on a map. This representation might not always be very straightforward to interpret, but we can colour-code it, too, by assigning colours for areas with the same concentration—e.g., from cold colours for the lowest values, through to warm colours for the highest values—in a way similar to the colour-coding of individual points (■ Fig. 4.20). This is one of the clearest ways to represent surface geochemical data, as our eyes are best suited for detecting

colours and patterns (and not numbers or letters, which require further processing by our mind). Thus, the colour-coded contour maps are the most used in geology and in geochemical exploration, as seen, for instance, in ■ Fig. 4.15. Sometimes this representation of the data is combined with colour-coding and size-coding of the individual dots representing the samples collected, or area colour coding can be used for one variable and dot colour-coding and size-coding for another variable. We are free to use the method that best suits our situation and our purpose, with the goal to convey the information—the position of the surface geochemical anomaly—quickly and reliably.

This brings us to another—very important—point: how exactly do we draw the lines of equal concentration? In other words, how can we reliably estimate the concentration of *any* point in the study area based on the *finite* number of data points available? We have to somehow extrapolate from 100 data points (in our example) to the whole of the 1 km^2 area. This is indeed a complicated and important task, and if not done reliably may lead to misestimating the size of the surface anomaly and of its position, leading to complications with further exploration efforts.

There are several methods of concentration estimation for an unknown point, the main of which are linear interpolation, inverse-distance interpolation, natural neighbours interpolation, and Kriging (■ Fig. 4.21); they all give different results, which is not very reassuring. Let us briefly consider each method and decide which one is preferable to use.

Linear interpolation is easy and straightforward to use, and relies on the assumption that the concentrations between two points of known values follow a linear relationship as a function of the distance between the points. Thus, knowing the distance between the two points and the values at the two locations, we can eas-

◘ Fig. 4.21 Concentration contours, representing a surface geochemical anomaly, calculated using four different methods, linear interpolation, inverse-distance interpolation, natural neighbours interpolation, and Kriging. The 0 values represent the local background, and all values above 0 represent the magnitude of the anomaly. The results are very different, both in terms of size and shape of the anomaly

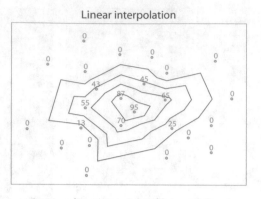

Linear interpolation

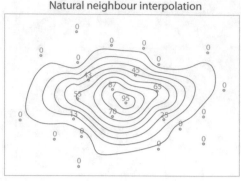

Natural neighbour interpolation

Inverse distance weighted interpolation

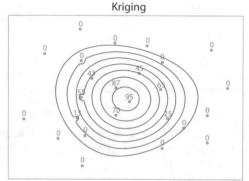

Kriging

ily calculate the concentration for any point in between. It is then straightforward to connect points of the same value. Another method, interpolation by triangulation, relies on the same assumption and gives the same results; mathematically the two are identical, even though the specific formulations are different.

The second method, inverse-distance interpolation, uses the distance between the point estimated and the available data points as a *weighing factor*: the farther the known point, the lower its participation in the estimation of the unknown point is. This method relies on the idea that the closer two points are to each other, the closer their concentrations will be to each other; the assumption is that at zero distance the concentrations will be the same, and at large distance they will be very different. The difficulty with this method is deciding how many points to use: the minimum is just the nearest three and the maximum is all points available. We do not have any independent factor or consideration allowing us to decide how many points to use; thus, geochemists are reduced to arbitrarily deciding on a number of known points to use or on the distance within which points are taken into consideration.

Natural neighbour is a somewhat complicated method, based on the intersection of polygons drawn with or without the estimated point. Firstly, Thiessen or Delauney polygons are drawn, with one data point per polygon; each of these polygons takes the known value. Then we draw new polygons, based on the point we want to estimate. The "old" polygons and the "new" polygons overlap to a certain degree: the new polygon is made of bits of the old polygons with different sizes. The con-

centration for the estimated point is calculated using the known values in the old polygons (only those participating in the new one), and on their proportion in the new polygon: the larger their proportion in the new polygon, the more weight is given to that value. Yes, the method is clunky and its logic not clearly apparent, yet it has its valiant and vigorous defenders.

When we consider these three methods of estimating values anywhere in the study area, we notice certain commonalities: they all rely on some basic assumptions. The major among these is that there is no concentration difference between samples collected immediately next to each other; in other words there is no zero-distance variability or uncertainty. The second commonality is that no pre-existent (background) geological variability is taken into account, and more specifically, it is assumed that there is no preferential background geochemical heterogeneity or orientation in the study area. Finally, we always use some sort of weighing: distance in linear and inverse-distance interpolations, and polygon overlap proportion in the natural neighbour method. This is, of course, a useful and logical approach; the only question is what reasoning is followed in assigning the weights and in deciding which the weighing factor will be. The three methods discussed above use purely geometrical weighing factors and arbitrarily assigned assumptions (e.g., linear lateral variability) and limitations (e.g., number of points to consider), without any attempt to provide some sound reason for these assumptions and limitations. This alone should disqualify these methods, also considering the extreme variability of results (◘ Fig. 4.21).

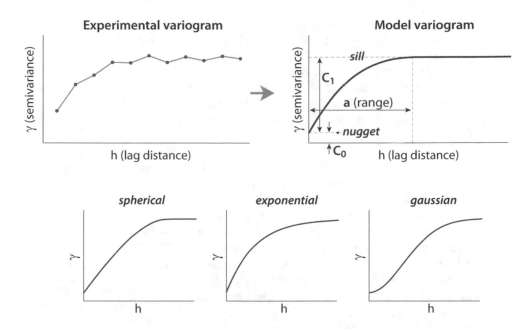

□ Fig. 4.22 Experimental and model variograms and their attributes. Examples of the most commonly used model variograms (spherical, exponential, gaussian) are also given

4.6.3 Kriging

The clear shortcomings of the examined interpolation methods were addressed in the 1960s by the South African D. Krige (after whom the method, *kriging*, is called), who challenged the accepted assumptions and sought to quantify the limitations. However, he accepted that any estimation of a point has to be based on the known points, the distance between the unknown point and the known points, and some sort of weighing factor. Most significantly, he provided a reliable method to quantify all the relevant factors, such as zero-distance variability (which he disagreed is nil), the number of points to consider (or limiting distance), and weighing factors. This method is the *variogram*.

The variogram is a simple curve in a variance versus lag distance space (□ Fig. 4.22), calculated using all the available data; the one based on our real-world samples is called the experimental variogram. Importantly, the specific relationship between the two (variance and lag distance) will be different for each project and must always be calculated anew. The calculation, in itself, is pretty simple:

$$\gamma = \frac{1}{2n} \sum \left(X_y - X_{y+h} \right)^2$$

where γ is the variance (or the semi-variance), n is the number of couples used, X_y is the concentration at a specific point y, and X_{y+h} is the concentration at a point situated at a specific distance (h) from the first one. The way this works is that we form pairs of all points separated by a specific distance (e.g., 100 m for our example in □ Fig. 4.19), then calculate the differences and square them, combine them and divide them by twice the number of pairs. The number we will obtain is the variance at lag distance of 100 m. Then, we repeat the calculation,

but take pairs situated at 200 m from each other: we thus have the variance at lag distance of 200 m. We continue on: take pairs 300 m apart and calculate the variance at lag distance of 300 m. For our example, we can calculate variances for all lag distances up to 1000 m. Then we plot the variances we calculated versus the lag distances used between the pairs (□ Fig. 4.22).

Once the experimental variogram is constructed, the best-fitting model variogram is found, with the most common being the spherical, exponential, and Gaussian ones. The attributes of the model variogram (□ Fig. 4.22) are directly used for the concentration calculation at any point:

– Firstly, we have to incorporate in our unknown point estimations the presence of a zero-distance variability; this is known as the *nugget effect*, named so as it is particularly strong in gold exploration, where the presence of a single gold nugget induces strong variability at very small distance. Thus, not only can we not assume that there is no variability at zero distance (as the other interpolation methods assume), but also we are able to estimate the size of this uncertainty. This residual uncertainty (C_0) explains the three points in the west on □ Fig. 4.21 not fitting the overall shape of the anomaly.

– Secondly, we are able to know how far we can go taking known points into consideration: the range (a) is the distance at which the variance has reached a high plateau (the sill; at $C_1 + C_0$): there is strictly no need—indeed it is *counterproductive*—to take into consideration any point beyond the range, as the variogram is informing us that the variance beyond the range is *not* a function of distance between points.

– Thirdly, we don't have to follow a linear model between points (in reality linear relationship between concentrations and distance is a rarity); rather we can

4

follow the equation describing the specific model variogram that best fits the experimental variogram in calculating the weights as a function of distance.

— Finally, we can construct experimental variograms in different directions (E–W, N–S, NE–SW, etc.), and appreciate any geochemical background orientation heterogeneity, which is not addressed by the other methods.

In conclusion, kriging not only successfully addresses the shortcomings of the most commonly used interpolations methods, but also provides us with a robust method to quantify the weighing factors and the distance to which points are taken into consideration, and to know the concentration–distance relationship. As such, and because it takes into account the pre-existing geochemical background orientation heterogeneity, kriging is the only method for constructing lines of equal concentrations that we should ever consider. Yes, there is some heavy-duty mathematics involved, but modern computers do the work for us in just a few clicks; this is also why kriging is now the standard method of point estimation and of building any lines of equal concentrations.

4.7 Summary

Geochemical exploration is likely the most common application of geochemistry. The general principles are that the element of interest (the commodity) typically migrates from a deposit at depth to the surface where it becomes trapped in a surface medium and forms a geochemical anomaly, which becomes our exploration target. The main ways of migration (or dispersion) generally fall in two groups, physical (e.g., erosion) and chemical (e.g., dissolution). The element of interest (and its accompanying pathfinder elements) can become trapped in a variety of surface media (soil, stream sediments, glacial till, vegetation, and water), and these are sampled and analysed to detect the surface anomaly.

Another major aspect of exploration geochemistry is the understanding of how and under what conditions deposits formed, specifically the formation pressure, temperature, and timing; the fluids involved and their chemistry; and the relevant processes.

Finally, the data obtained have to be interpreted as a function of a variety of environmental factors and visualized in a reliable and logical way. The best geostatistical method, taking into account the spatial variability of the element analysed, is kriging.

❓ Exercises

Q4.1 What is a geochemical anomaly?

Q4.2 What are the major dispersion modes?

Q4.3 How do environmental factors and Eh and pH affect the mobility of elements?

Q4.4 What is the importance of isotopes in the study of mineral deposits?

Q4.5 What are the four main surface sampling media, when bedrock is not exposed?

Q4.6 Which soil horizon is the most useful in exploration, and why?

Q4.7 Using the data from ◻ Fig. 4.19, construct the E–W, N–S and NE–SW variograms. For each orientation, quantify the nugget effect, the sill, and the range. Comment on any differences between them.

References

1. Alexandre P (2010) Mineralogy and geochemistry of the Aricheng Na metasomatism-type uranium occurrence, Guyana. Miner Deposita 45:351–367
2. Alexandre P (2020) Isotopes and the natural environment. Springer, Berlin, 97 pp
3. Alexandre P, Kyser K, Thomas D, Polito P, Marlat J (2009) Geochronology of unconformity-related uranium deposits in the Athabasca Basin, Saskatchewan, Canada and their integration in the evolution of the basin. Miner Deposita 44:41–59
4. Alexandre P, Heine T, Fayek M, Potter E, Sharpe R (2019) Ore mineralogy of the Chisel Lake Zn-Cu-Ag (+Au) VMS deposit in the Flin Flon-Snow Lake Domain, Manitoba, Canada. Can Mineral 57:925–945
5. Baas-Becking LGM, Kaplan IR, Moore D (1960) Limits of the natural environment in terms of pH and oxidation-reduction potentials. J Geol 68:243–284
6. Bailey JC (2006) Geochemistry of boron in the Ilímaussaq alkaline complex, South Greenland. Lithos 91:319–330
7. Cathelineau M (1988) Cation site occupancy in chlorites and illites as a function of temperature. Clay Miner 23:471–485
8. Cathelineau M, Izquierdo G (1988) Temperature-composition relationships of authigenic micaceous minerals in the Los Azufres geothermal system. Contrib Miner Petrol 100:418–428
9. Cathelineau M, Nieva D (1985) A chlorite solid solution geothermometer. The Los Azufres geothermal system (Mexico). Contrib Miner Petrol 91:49–57
10. Clausen FL (1982) A geostatistical study of the uranium deposit at Kvanefjeld, the Ilimaussaq Intrusion, South Greenland. Risø National Laboratory, Roskilde, Denmark, report Risø-R-468, 289 p
11. Criss RE, Singleton MJ, Champion DE (2000) Three-dimensional oxygen isotope imaging of convective fluid flow around the Big Bonanza, Comstock Lode Mining District, Nevada. Econ Geol 95:131–142
12. de Caritat P, McPhail DC, Kyser K, Oates CJ (2009) Using groundwater chemical and isotopic composition in the search for base metal deposits: hydrogeochemical investigations in the Hinta and Kayar Pb–Zn districts, India. Geochem A Exploration Environ Anal 9:215–226
13. Hutchinson MN, Scott SD (1988) Sphalerite geobarometry applied to metamorphosed sulfide ores of the Swedish Caledonudes and US Appalachians. Norges Geol Unders Bull 360:59–71
14. Isaacs EH, Srivastva RM (1989) An introduction to applied geostatistics. Oxford University Press, 561 pp
15. Kyser TK (1987) Equilibrium fractionation factors for stable isotopes. In: Kyser TK (ed) Stable isotope geochemistry of low temperature processes, mineralogical association of Canada short course handbook, vol 13
16. Ossandon G, Freraut R, Gustafson LB, Lindsay DD, Zentilli M (2001) Geology of the Chuquicamata mine: a progress report. Econ Geol 96:249–270

17. Polito P, Kyser K, Lawie D, Cook S, Oates C (2007) Application of sulphur isotopes to discriminate Cu–Zn VHMS mineralization from barren Fe sul-phide mineralization in the greenschist to granulite facies Flin Flon-Snow Lake–Hargrave River region, Manitoba, Canada. Geochem Exploration Environ Anal 7:129–138

18. Sørensen H, Rose-Hansen J, Nielsen BL, Løvborg L, Sørensen E, Lundgaard T (1974) The uranium deposit at Kvanefjeld, the Ilímaussaq intrusion, South Greenland. Rapport Grønlands Geologiske Undersøgelse 60, 54 pp

19. Thomas JB, Watson EB, Spear FS, Shemella PT, Nayak SK, Lanzirotti A (2010) TitaniQ under pressure: the effects of pressure and temperature on the solubility of Ti in quartz. Contrib Min Petrol 160:743–759

20. Urqueta E, Kyser TK, Clark AH, Stanley CR, Oates CJ (2009) Lithogeochemistry of the Collahuasi porphyry Cu–Mo and epithermal Cu–Ag (–Au) cluster, northern Chile: pearce element ratio vectors to ore. Geochem Exploration Environ Anal 9:9–17

21. Wark DA, Watson EB (2006) TitaniQ: a titanium-in-quartz geothermometer. Contrib Min Petrol 152:743–754

Further Reading

22. **Introduction to Exploration Geochemistry**, 2nd edition, by A. A. Levinson (1980) Applied Publishing Ltd., Wilmette, IL, USA. ISBN 0-915834-04-9.

23. **Biogeochemistry in Mineral Exploration**, by C. Dunn (2007) Elsevier Science, ISBN: 978-0-444-53074-5.

24. **Fluid Inclusions: Analysis and Interpretation**, by I. Samson, A. Anderson, and D. D. Marshall (2003) Mineralogical Association of Canada, ISBN: 978-0-921-29432-0.

25. **Fluid inclusions**, by E. Roedder (1984) Volume 12 of Reviews in Mineralogy, Mineralogical Society of America, Walter de Gruyter GmbH & Co KG, ISBN 978-1-501-50827-1.

26. **Isotopes and the Natural Environment**, by P. Alexandre (2020). Springer, ISBN: 978-3-030-33651-6.

27. **An Introduction to Applied Geostatistics**, by E.H Isaaks and R.M. Srivastva (1989) Oxford University Press, ISBN: 0-19-50512-6.

Environmental Geochemistry

Contents

5.1   The Significance of Environmental Geochemistry – 86
5.1.1   The Anthroposphere – 86
5.1.2   Applications of Environmental Geochemistry – 86

5.2   Past Climate Change – 87
5.2.1   The Historic Compilation of Guy Callendar – 87
5.2.2   Climate Change Causes – 88
5.2.3   Records of Paleo-Temperatures – 88
5.2.4   Isotopes as Paleo-Temperature Proxies – 89

5.3   Tracing Industrial and Agricultural Pollution – 90
5.3.1   General Principles of Pollution Tracing – 90
5.3.2   Lead as Industrial Pollution Tracer – 90
5.3.3   Nitrogen and Water Pollution – 92

5.4   Environmental Biogeochemistry and Remediation – 94
5.4.1   Definition of Biogeochemistry – 94
5.4.2   Environmental Biogeochemistry and Remediation – 94
5.4.3   Adsorption and Remediation – 96

5.5   Summary – 98

References – 98

Electronic supplementary material The online version of this chapter (▶ https://doi.org/10.1007/978-3-030-72453-5_5) contains supplementary material, which is available to authorized users.

© Springer Nature Switzerland AG 2021
P. Alexandre, *Practical Geochemistry*,
Springer Textbooks in Earth Sciences, Geography and Environment,
https://doi.org/10.1007/978-3-030-72453-5_5

5.1 The Significance of Environmental Geochemistry

5.1.1 The Anthroposphere

Compared to other animal species, we humans have very highly developed technology, communication, transportation, education, and a very complex society. Unsurprisingly, this extraordinary development has made us believe that we are also somehow exceptional, to the point of considering ourselves the "kings of creation" to whom everything else is subordinated. We now effectively live in the *anthroposphere* [9]: an entirely human-generated environment, which itself has complex relationships with the other environments or components of the Earth (geosphere, biosphere, atmosphere, and hydrosphere; ● Fig. 5.1). The anthroposphere has its own components, such as technology (or the technosphere) including the internet; resource and energy extraction (mining, oil, farming, fishing, and forestry); domesticated species; urban development and infrastructure; and societies with their own complex internal and external interactions.

One of the crucial considerations here is our—the human species'—position, role, and interactions with the natural environment. We are a very invasive species, much more than any other that has existed before us, or currently exists. We aggressively use any natural material as a resource for our purposes, as we tend to have the attitude that we are entitled to Earth's resources without any accountability to anyone; our complex technology

and expansive lifestyle have led us to utilize any part of the Earth as resource supply for us. If you look at the periodic table of the elements (● Fig. 1.3), you will not be able to find an element that is not also a *commodity*, a resource for us; the same applies to pretty much any object, entity, or item present on Earth. This includes the entirety of the other Earth components (geosphere, biosphere, atmosphere, and hydrosphere). The net result is two-fold:

— On the one hand, our way of life and the entire anthroposphere are totally and completely unsustainable. While the biosphere has a complex set of processes and mechanisms to produce and recycle materials (e.g., photosynthesis and decomposition) and remains in a global equilibrium, the anthroposphere is highly inefficient at sustaining itself: we extract resources at an ever-increasing rate, but have no mechanisms to replace them in the natural environment.

— On the other hand, we constantly damage the Earth, in two ways: our extraction of resources is done in a way that is very damaging to the natural environment; and all our activities generate waste and pollution that profoundly affect the biosphere, atmosphere, and hydrosphere (and to a lesser extent, the geosphere). This has come to the point that some chemical elements (e.g., arsenic, mercury, lead) can now be defined as *biophile* elements [7].

We—all of us sentient humans—have to be fully aware of the effects we have on the natural environment. We have to significantly change our attitude, from "I'm the boss and I do what I want" to "We're all in this together and have to take care of each other." We have to find ways to be far less invasive and damaging to the natural environment, before, hopefully, finding ways to reverse some of the damage we have done, before it's too late.

5.1.2 Applications of Environmental Geochemistry

There are plenty of people, a small army of scientists in a variety of institutes and organizations, who work hard in the direction of reversing humans' negative impact on the Earth. Simplistically, there are two main axes of environmental research: the theoretical approach and the practical one. On the side of theory, we try to understand the natural environment (e.g., how it works, what processes occur, what factors affect it) and study the changes and modifications to the natural environment (both natural and human-induced) and what factors affect these changes. On the side of practical applications, we work primarily on detecting and remediating industrial and agricultural pollution. And geochemistry is a powerful tool that can be of significant help in environmental research, as it gives us the theoretical and

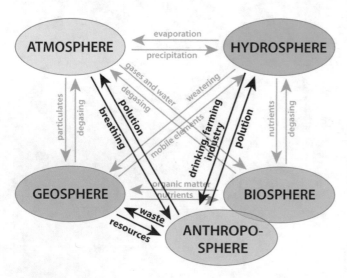

● **Fig. 5.1** Position of the anthroposphere relative to the other components of the Earth (geosphere, atmosphere, hydrosphere, and biosphere), and summary interactions between them. Given that humans are animals, and considering domesticated species and farming, there is an overlap between the anthroposphere and the biosphere. Crucially, humans utilize resources from all other spheres, leaving them severely damaged and polluted. After Kuhn and Heckelei [9], modified

practical tools to take care of our natural environment by reducing our negative impact on it.

In a way, environmental geochemistry is akin to geochemical exploration, as some of the same basic principles and premises apply to both. In exploration, we try to find a surface anomaly as an expression of a hidden deposit; in environmental studies, we try to trace a surface pollution to its origin. In both cases, we are very interested in how the element of interest (a commodity in exploration, a pollutant in environmental studies) migrates in the near surface environment. In both fields, we are keenly interested in understanding the processes and factors at play, be it for the formation of a deposit, or for the formation of a polluted environment. In both cases, we incorporate our knowledge and understanding of how natural systems work in any part of the Earth (geosphere, atmosphere, hydrosphere, and biosphere, but also the anthroposphere).

The applications of environmental geochemistry are many and varied. Many focus on the theoretical understanding of the environment, which we will examine here. We will also see a few examples from the practical side of environmental geochemistry, such as climate change, industrial and agricultural pollution, biogeochemistry, and remediation studies.

5.2 Past Climate Change

The notion of climate change is not new. As far back as in Ancient Greece, philosophers like Aristotle and his disciple Theophrastus were aware that climate changes and proposed sound theories as to the cause of the change, some of which hold true today. With the advent of the scientific revolution in the sixteenth to eighteenth centuries, people like Charles de Montesquieu and Jean-Baptiste Dubos of France, David Hume of England, and Thomas Jefferson (the third president of the United States) were commenting on the change in climate and the effects of this change on society. With the strong development of science in the early nineteenth century, scientists John Tyndall and Svante Arrhenius—among others—started speculating about the relationships between the concentrations of gases in the atmosphere (H_2O, CO_2, O_3, CH_4) and its temperature. They were able to lay the foundation of the science behind the greenhouse effect and climate change. These efforts then culminated in the first specific and concrete factual evidence of climate change, provided by a certain Guy Callendar.

5.2.1 The Historic Compilation of Guy Callendar

Guy Callendar was an engineer and inventor, but also an amateur climatologist, as we would say today. Through his government work, he had access to an extensive data set of temperature measurements from around the world and was able to compile them to demonstrate that the air temperature was indeed rising at a steeper rate than in the pre-industrial era (Fig. 5.2).

Callendar concluded that the overall atmospheric temperature had risen by nearly 0.5 °C between 1890 and 1935. He also claimed that emissions of CO_2 from burning fossil fuel (mostly coal) could cause a "greenhouse effect" and that this was the cause of the warming. Building on Arrhenius' work, Callendar asserted that increasing CO_2 concentration in the atmosphere resulted in higher retention of thermal energy and thus caused increasing temperatures. This theory was called the Callendar effect, and its predictions were later proven to have been remarkably accurate when climate science developed in the 1950s and 1960s. Later, in 1941, Callendar published another paper explaining how the increase of CO_2 concentrations—and also those of other "greenhouse" gases such as N_2O and CH_4—caused an increase in temperature.

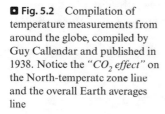 **Fig. 5.2** Compilation of temperature measurements from around the globe, compiled by Guy Callendar and published in 1938. Notice the *"CO_2 effect"* on the North-temperate zone line and the overall Earth averages line

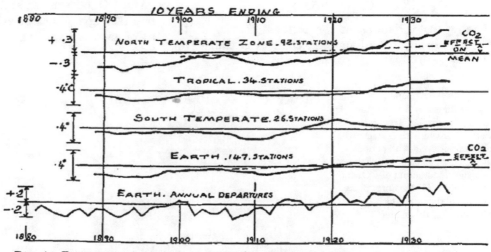

FIG. 4.—Temperature variations of the zones and of the earth. Ten-year moving departures from the mean. 1901-1930. °C.

5

5.2.2 Climate Change Causes

While Calendar focused on the increase of "greenhouse" gases as the cause of rising atmospheric temperatures, there are also other causes at work. On the planetary scale and in the long term, there are two main groups of causes for temperature variations on the Earth's surface, and these are generally grouped as *astronomical* and *terrestrial*. The former is controlled by the slight variations of Earth orbit's ellipticity and rotational axis inclination; they are entirely predictable (according to the Milankovitch cycles) and have periods of 19, 22, 24 thousand years (for the precession), 41 thousand years (for the obliquity), and 95, 125, and 400 thousand years (for the eccentricity). To these we can add the variations in solar activity, which are less cyclic and therefore less predictable.

The terrestrial causes are more complex and even less predictable, as they depend on two forcing feedbacks working in opposite directions (▪ Fig. 5.3). The first is called the *albedo* feedback: as we have more snow and ice on earth's surface, more and more of sun's heat will be reflected (high albedo), causing further cooling of the atmosphere. It is when this feedback had reached extreme proportions that the Earth was covered in ice from pole to pole (the "Snowball Earth" episodes). The second is called water vapour feedback: when air temperatures increase, more evaporation occurs resulting in an elevated greenhouse effect and higher air temperatures. If this feedback is left alone, it can accelerate and result in high air temperatures with no ice, including on the poles. Indeed, more than half of Earth's history was ice-free, as is demonstrated by the sedimentary record.

However, there is also a third feedback, a moderating—or stabilizing—one, which involves erosion and the weathering of rocks on the continents, and thus a decrease of CO_2 in the atmosphere, through the Urey reaction:

$$CaSiO_3 + CO_2 = CaCO_3 + SiO_2$$

Finally, the variations in the tectonic activity of the Earth also affect the long-term changes in climate: we have periods of increased volcanic activity and release of more CO_2 and other volatiles in the atmosphere, and subduction, which results in recycling of carbon and decreasing temperatures.

Of course, there are several other human activities that affect climate beside the burning of fossil fuel. These include—but are not limited to—emissions of aerosols, cement manufacturing, deforestation, livestock breeding, use of fertilizers, industrial activity; the list is long. Human activities are, however, the only climate change factor that we are in control of; they also heavily outweigh all other causes of atmospheric temperature variations.

5.2.3 Records of Paleo-Temperatures

The relationship between the amount of greenhouse gases—and in particular of CO_2—in the air and atmosphere's temperature has been long established (e.g., [18]) and shows a direct relationship between anthropogenic emissions and temperature, with the other—natural—causes of temperature increase mentioned earlier having negligible effect.

The evidence of past climate change can be found in the sedimentary and glacial record. For instance, ancient ice found in the circumpolar regions contains tiny bubbles of air trapped during ice formation. It is possible to drill through the ice and extract drill core samples from a specific time period. The trapped air is extracted and its CO_2 concentration is measured and compared with an independent temperature record (▪ Fig. 5.5). Independent measures of paleo-temperatures include biological information (analysis of pollen and beetle distribution, for instance), historical records (such as Callendar's compilation), but also tree records (the information recorded in tree rings, each corresponding to a year), the glacial record (studying ancient ice), and the geological record. In that manner, the air temperature dependence on air CO_2 concentration that was observed in the recent past (▪ Fig. 5.4) can be extended as far as the availability of ancient ice allows.

▪ **Fig. 5.3** Schematic representation of the two main forcing feedbacks affecting Earth's climate. These do involve the tectonic cycle, with periods of increased volcanic activity and thus increased output of CO_2 and other volatiles in the atmosphere, and of subduction, which recycles carbon in Earth's mantle

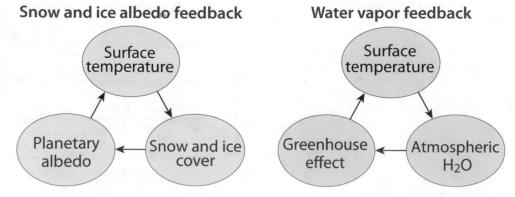

Snow and ice albedo feedback

Surface temperature → Snow and ice cover → Planetary albedo → Surface temperature

Water vapor feedback

Surface temperature → Atmospheric H_2O → Greenhouse effect → Surface temperature

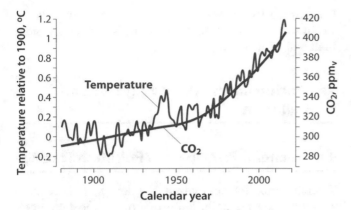

◙ Fig. 5.4 Evolution of the mean annual surface temperature, in red, and the concentration of CO_2 in the atmosphere, in blue, between 1881 and 2017, showing a clear relationship between the two. Data sources National Aeronautics and Space Administration, National Oceanic and Atmospheric Administration, and Goddard Institute for Space Studies

5.2.4 Isotopes as Paleo-Temperature Proxies

Another powerful proxy for tracking paleo-climate variations is the oxygen isotopic record in sedimentary rocks, in particular carbonates. The mechanism through which carbonate rocks reflect the atmosphere temperature is the following: during a colder period, the amount of ice in circumpolar regions increases, leading to a higher degree of sequestration of ^{16}O (the oxygen isotope most represented in circumpolar ice sheets), resulting in lower amount of available ^{16}O and thus higher $\delta^{18}O$ values of the remaining free ocean water. These higher $\delta^{18}O$ values will be reflected in the oxygen isotopic composition of biogenic carbonates (foraminifera, molluscs, ostracodes, ooliths, stromatolites, etc.), biogenic phosphates

(teeth or bone), silica (cherts, diatoms), alunite, biogenic cellulose, or clay minerals (kaolinite and smectite). In other words, there is a clear relationship between water oxygen isotopic composition and global air temperature (◙ Fig. 5.6): warmer, interglacial conditions are reflected in lower $\delta^{18}O$ values of carbonates and colder periods correspond to higher $\delta^{18}O$ values of carbonates. As a rule of thumb, a decline of 1‰ of $\delta^{18}O$ typically reflects about 1.5 °C warming.

Another isotopic tracer of past climate change are the $^{87}Sr/^{86}Sr$ ratios of carbonate sedimentary rocks. High $^{87}Sr/^{86}Sr$ ratios are observed in continental rocks as they correspond to higher degree of differentiation of igneous rocks (▶ Chap. 1). When the climate is colder and more ice is formed in circumpolar regions, the global ocean level will decrease, leading to higher degrees of physical erosion of continental rocks, resulting in higher amounts of Sr with high $^{87}Sr/^{86}Sr$ ratios being mobilized and transported to the ocean. Thus, during colder periods ocean water has higher $^{87}Sr/^{86}Sr$ ratios, recorded in the same minerals used to analyse oxygen isotopic composition, listed above. The two isotopic proxies, $\delta^{18}O$ value and $^{87}Sr/^{86}Sr$ ratios of carbonates, work in opposite directions, but can be used in conjunction with each other to derive paleo-temperatures [14, 19].

Using a variety of independent methods and the dependence of the extent of fractionation on temperature (see ▶ Chap. 1; [2]), the actual temperature of ocean water can be estimated, with a good degree of certainty for a sample of carbonate sediment, specifically those containing benthic foraminifera (◙ Fig. 5.6). Of course, several uncertainties (fractionation factors, ancient ocean water isotopic compositions, kinetic fractionation) as well as diagenetic processes, contribute to make the temperature estimates less certain.

◙ Fig. 5.5 Temperature record in the Vostok ice core, Antarctica, showing strong correlation to CO_2 concentrations (after Petit et al. [15], modified)

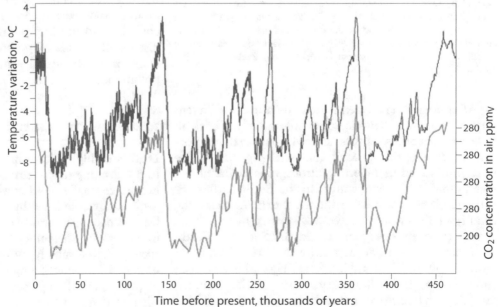

5

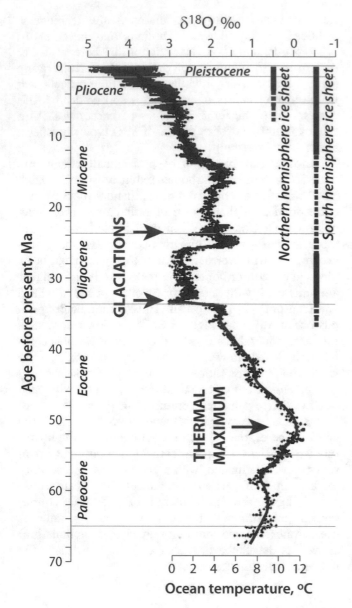

Fig. 5.6 Oxygen isotopes in foraminifera, reflecting significant changes in ocean temperatures over the last 70 million years. Strong increases in δ¹⁸O correspond to the presence of ice sheets in the circumpolar regions sequestrating ¹⁶O: specific glaciation events can be identified on the basis of oxygen isotopes. From Zachos et al. [20], modified

After compiling oxygen isotope data in the sedimentary record, from benthic foraminifera, as far back in time as possible, scientists noticed some extreme positive excursions of the δ¹⁸O values (Fig. 5.6). These were interpreted as reflecting strong dips in the global mean atmospheric temperature at the Earth' surface, or extreme glacial periods. In the most extreme cases, the totality of the earth's surface was covered by ice, which is why these episodes were called the Snowball Earth. This likely occurred at least twice in Earth's history, at around 650 and 700 million years ago. This has been confirmed by the sedimentary record: glacial sedimentary deposits have been found in equatorial regions.

Contrasting to the Snowball Earth is an ice-free Earth, which was the condition of the Earth's surface for more than half of its history.

5.3 Tracing Industrial and Agricultural Pollution

5.3.1 General Principles of Pollution Tracing

Tracking, monitoring, quantifying, and helping remediate anthropogenic pollution is one of the most common practical applications of environmental geochemistry. As noted earlier, virtually any human activity has detrimental effect on the environment, mostly by destroying habitat, changing the environment, and emitting pollutants. Here, we will consider how geochemical methods can help trace the source of both industrial and agricultural pollution.

The methods used to trace the source of pollution are fairly similar to those used in exploration geochemistry, discussed in ▶ Chap. 4, and follow the same general principle. Basically, we try to detect anomalous concentrations trapped in surface media (water, soil, vegetation) and use them to find the source of these anomalous concentrations. (In exploration, the surface anomaly will lead us to the deposit at depth.) Crucially, the same rules and factors controlling the mobility and trapping of elements in surface media, discussed in ▶ Chaps. 1 and 4 and used in exploration, will be fully applicable when tracing anthropogenic pollution.

As mentioned, any human activity generates waste and causes pollution. Any industrial process (e.g., mining, smelters, metal and oil refineries, chemistry, pharmaceuticals, and many others) and many agricultural practices (e.g., fertilizing, intensive animal husbandry), but also our everyday life (e.g., domestic waste, burning of fossil fuels, the funeral industry, to name a few), all contribute to produce pollution. We will not be able to discuss all sources of pollution here, of course, but will consider a few typical examples, starting by lead (Pb) pollution.

5.3.2 Lead as Industrial Pollution Tracer

Lead is highly toxic to humans and animals—it causes lead poisoning (a.k.a. saturnism)—and is also prevalent in surface media. As far as 6,000 years ago, anthropogenic lead was emitted by mining and smelting of silver, lead, and copper ores. The process accelerated sharply in the eighteenth century by the industrialization of our society, driven mostly by mining, smelting, but also by burning of Pb-containing coal (which led to its own set of pollutions). Things became much worse in the 1920s, when lead stared to be added into gasoline, in the form

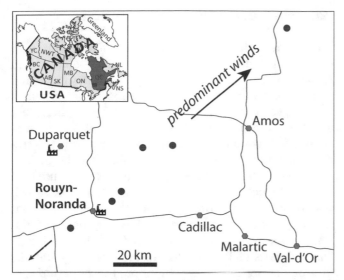

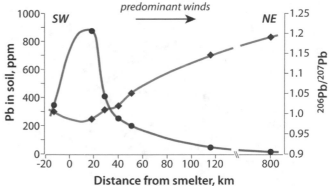

Fig. 5.7 Position of the Horne copper smelter in Rouyn–Noranda (Quebec, Canada) and of the soil sampling sites (red dots). The predominant winds here are to the NE, with a small proportion to the SW. From Hou et al. [8], modified

Fig. 5.8 Pb concentrations (in ppm) and ^{206}Pb/^{207}Pb ratios in soils at varying distances upwind and downwind from the smelter. Background levels are reached only beyond 100 km away from the smelter. From Hou et al. [8], modified

of antiknock alkyl lead-based additives; lead pollution from internal combustion engine exhaust now exceeds all other sources of anthropogenic lead emissions. All these anthropogenic Pb pollution sources combined account for more than 95% of Pb found in the surface environment.

We can measure Pb in surface media such as natural waters (streams and lakes), in plants (bark, wood, and leaves), and in soil. As an example, soils in vicinity of the Horne copper smelter in Rouyn–Noranda (Quebec, Canada)—a heavily polluted mining and smelting area—were analysed as a function of distance to the smelter ([8]; ■ Fig. 5.7). The predominant winds here are to the northeast, transporting the air-borne Pb to very significant distances.

Lead concentrations and lead isotopes were analysed in 75 soils in seven sampling sites (6 downwind, one upwind), distant up to 120 km away, with one sample 800 km away downwind. Strong Pb enrichments were observed in soils several kilometres downwind from the smelter (■ Fig. 5.8), ranging from 300 to 1000 ppm, and background levels were reached only farther than 100 km from the smelter.

In this situation, lead isotopes vary between the anthropogenic—smelter-related—contribution (^{206}Pb/^{207}Pb ratios of 0.95–1, equivalent to those of the mined sulphide ore) and the local background level, corresponding to the local rocks (^{206}Pb/^{207}Pb ratios of 1.15–1.4; ■ Fig. 5.8). The isotopic compositions of pollutants—Pb from a smelter in this case—will often be very different from those of natural sources and will allow us to detect and visualize the extent of the pollution even when the actual concentrations are very low. In this case, Pb concentrations reach background

levels shortly after 100 km away from the smelter, but the ^{206}Pb/^{207}Pb ratio is still noticeably different even significantly further away (■ Fig. 5.8).

We can also use the isotopic ratios to distinguish between the smelter-related contribution to pollution from other anthropogenic sources, notably from lead additives in fuel. The ^{206}Pb/^{207}Pb ratio of Rouyn–Noranda ore and of local lithologies is known, as shown above, and distinct from that of leaded fuel. Using these ratios and the isotopic mixture calculations described in ▶ Chap. 3, it is possible to estimate the smelter's contribution to pollution to approximately 85% near the smelter to approximately 50% away from it, the rest coming mostly from leaded gasoline.

Samples were collected from different depths in the soil at all sampling sites, resulting in depth profiles showing decrease in Pb concentrations and increase in ^{206}Pb/^{207}Pb ratios with depth, tending to the local background values (■ Fig. 5.9). These profiles demonstrate that the polluting lead was transported in the air, most likely as particles, and not by groundwater: had it been transported by groundwater, the values would have been much more uniform as function of depth.

This example demonstrates that geochemical methods, very similar indeed to those used in mineral exploration, are a very powerful tool in detecting, tracing, and quantifying industrial pollution. We can have a very clear idea of the source of the pollution, its extent (both in terms of distance and amounts), and the possible transport mechanisms, which are some very useful pieces of information that can be used in remediation. We can apply this method to other pollutants (e.g., Cd, Zn, Ni, and Cu), using both concentrations and isotopic ration and a variety of surface sampling media (soil, vegetation, water, but also air). Something very similar is true for agricultural pollution, and this is the next example of detecting and tracing pollution that we will consider.

5

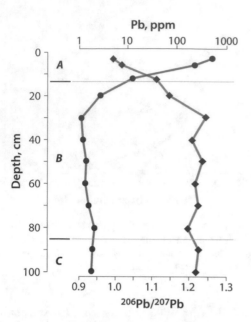

☐ **Fig. 5.9** Depth profile of Pb concentration and $^{206}Pb/^{207}Pb$ ratio in soil, from a site approximately 14 km downwind from the smelter in Rouyn–Noranda. The airborne character of the pollution is clearly demonstrated by the shape of the profiles, with the smelter-related contribution (high Pb concentrations, low $^{206}Pb/^{207}Pb$ ratios) restricted to the A horizon and the top of the B horizon. Background levels are reached approximately 30 cm below surface. From Hou et al. [8], modified

5.3.3 Nitrogen and Water Pollution

Water is very significant for any form of life, as it is an essential nutrition necessity for plants, wildlife, and humans alike. It is part of the global water cycle and interacts extensively with all other Earth systems, such as the geosphere, the atmosphere, the biosphere, but also the anthroposphere. Water that is most directly useful to life is contained in groundwater, which is the second most voluminous water reservoir, after the oceans. Drinking water consumed by humans and used in agriculture comes from groundwater's aquifers, which underlines the importance of healthy groundwater for farming, humans, and animals. However, most groundwater is severely polluted (not only in the heavily industrialized parts of the world). The amounts of pollutants in groundwater are strictly monitored in most jurisdictions, in particular for drinking—both bottled and tap—water, (☐ Table 5.1 and ☐ Fig. 5.10).

One of the most dangerous components of groundwater is nitrogen. In small amounts, it is nutrition to plants, but in elevated concentrations it is toxic to animals and humans. It exists in two main forms in groundwater, nitrate (NO_3^-) and other nitrous oxides (NO_X, collectively known as "noxes"), and ammonium (NH_4^+). Nitrate is toxic to humans and most jurisdictions define the safe concentrations at 10 mg/L; ammonium is toxic

☐ **Table 5.1**   Concentrations of dissolved ions in a small selection of bottled waters, mostly from Europe (where such information is compulsory on the labels). Fluorides, pH, and some toxic metals (e.g., As, Pb; see ☐ Fig. 5.10) are sometimes listed. Note that nitrates are almost always within the mandated limits

Bottled waters	Dissolved ion concentrations, mg/L							
	Ca^{2+}	Mg^{2+}	Na^+	K^+	Cl^-	NO_3^-	SO_4^{2-}	HCO_3^-
Ashbeck, UK	10.0	2.5	9.0	2.0	12.0	11.0	10.0	25.0
Bonpreau, Spain	50.7	7.5	5.3	0.7	4.9	5.1	32.4	154.0
Buxton, UK	55.0	19.0	24.0	1.0	37.0	0.1	13.0	248.0
Canadian Spring, Canada	35.0	7.0	4.0	2.0	4.0	0.0	20.0	139.0
Carrefour, France	4.1	1.7	2.7	0.9	0.9	0.8	1.1	25.8
Evian, France	78.0	24.0	5.0	1.0	4.5	3.5	10.0	357.0
Fontenoche, Italy	16.8	4.9	6.1	1.7	7.3	2.0	8.2	83.6
Speyside Glenlivet, UK	12.0	1.6	3.9	0.7	5.0	1.0	4.0	120.0
Volvic, France	11.5	8.0	11.6	6.2	13.5	6.3	8.1	71.0
Vittel, France	91.0	19.9	7.3	0.0	0.0	0.6	105.0	258.0
Waitrose, UK	18.0	4.5	12.0	0.8	26.0	7.8	17.5	39.0
Averages								
11 waters listed above	34.7	9.1	8.3	1.5	10.5	3.5	20.8	138.2
1887 mineral waters[a]	41.6	12.4	4.0	2.3	11.3			217.0
3317 bottled waters[a]	38.3	11.3	2.0	2.3	12.2			205.0

[a]Data from openfoodfacts.org

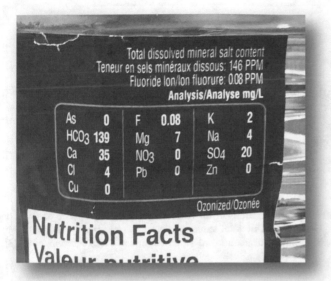

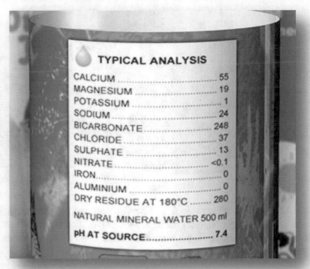

☐ Fig. 5.10 Concentrations of dissolved cations (Ca²⁺, Mg²⁺, Na⁺, and K⁺), oxyanions (HCO₃⁻, SO₄²⁻, Cl⁻, and NO₃⁻), and common toxic elements (Pb, As, Cu, and Zn), listed on the labels of two bottled natural mineral waters. "Noxes", or the various nitrous oxides, are highly toxic to humans and are therefore strictly monitored. In this example, noxes are well within the safe limit, which for most countries is 10 mg/L

to humans and to aquatic life, and the safe limit is most commonly set at 1 mg/L.

Simplistically, there are three major sources of nitrogen in groundwater: the natural decay of biomass, fertilizers, and manure or septic effluents. It is sometimes complicated to distinguish which of the two pollutants, fertilizers or manure, is the dominant one. In this case, stable nitrogen isotopes are very helpful: the different sources have distinct isotopic compositions (☐ Fig. 5.11). In particular, fertilizers and manure have clearly different $\delta^{15}N$ values: manure has higher δ^{15} N, which is due to the metabolic effects during digestion of plant food (e.g., grass, hay, feed; [2]).

One example of isotopes providing the clue to the source of agricultural pollution comes from south

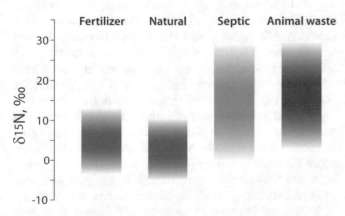

☐ Fig. 5.11 Nitrogen isotopic compositions of different sources. Fertilizers and natural organic matter in soil have similar $\delta^{15}N$ values, at approximately 0–10‰, whereas septic effluents and manure (which are effectively the same thing) have distinct composition of approximately 5–25‰

Ontario, Canada, where a very significant nitrates pollution was observed near the city of Cambridge. The nitrate concentration was very high, at approximately 160 mg/L [17], or 16 times the generally accepted safe level. Of course, the source of this pollution had to be traced, but the concentrations in themselves were fairly homogeneous and did not provide any useful information as to the pollution source. However, stable oxygen and nitrogen isotopes were measured in groundwater and proved to be very valuable. Several wells were drilled, and groundwater was sampled at various depths below the water table, resulting in an isotopic cross-section through the polluted area (☐ Fig. 5.12). Oxygen isotope values were fairly similar through the area and did not distinguish the two pollution sources. However, nitrogen isotope values detected two distinct groundwater pollution populations, one corresponding to fertilizers ($\delta^{15}N$ values at approximately 4.6‰) and another corresponding to sceptic effluents ($\delta^{15}N$ values above 5‰; [3]). Knowing the local ground flow direction allowed scientists to retrace the septic-related pollution to some abandoned and previously unknown septic tile drains, where a cluster of long forgotten farmhouses once stood (☐ Fig. 5.12). This example is one of many similar cases, where pollution was traced to its source using not only concentrations but also stable isotopes, and in particular nitrogen isotopes (e.g., Corcoran [4]).

As mentioned earlier, environmental geochemistry borrows heavily from exploration geochemistry. This example will remind us the case of the Hinta Pb–Zn prospect in Rajasthan, India (discussed in ▶ Chap. 4), where sulphur isotopes measured in groundwater reflected the presence of a hidden deposit and it was pos-

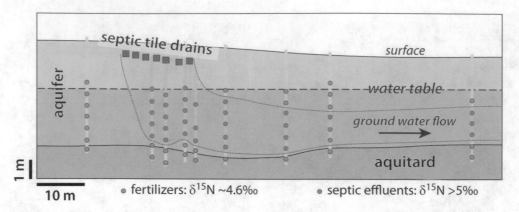

5

□ **Fig. 5.12**   Cross section through soil heavily polluted by nitrates, with NO_3^- concentrations of approximately 160 mg/L [17]. The position of the wells and the water samples taken are indicated. The septic effluents, characterized by $\delta^{15}N$ values of above 5‰, are clearly different from fertilizer-related pollution, which permitted researchers to trace the specific pollution to abandoned septic tile drains. After Aravena et al. [3], modified

sible to find its approximate location knowing the local groundwater flow direction. We can therefore conclude that geochemical methods not dissimilar to those used in exploration—measuring elemental concentrations and isotopic values in a variety of surface sampling media—provide a very powerful method to discover the source of industrial, agricultural, or other pollution, to quantify its extent in terms of space and concentrations, and to have an idea of the pollutant dispersion mechanisms.

5.4   Environmental Biogeochemistry and Remediation

5.4.1   Definition of Biogeochemistry

In the most general terms, biogeochemistry is the branch of science dealing with the interactions between the rock realm (the geosphere) and the realm of living beings (the biosphere), approached from a chemical point of view. In practice, things become more complicated as other Earth components, in particular the hydrosphere and the atmosphere, become involved. By its nature, biogeochemistry is often concerned with soil, which is the environment situated at the geosphere–biosphere interface, but also involves the hydrosphere and the atmosphere. Biogeochemistry often deals with cycling of major chemical elements present in all Earth's components, primarily oxygen, carbon, and nitrogen, and their interactions with living beings, and with the processes involving the geosphere and the biosphere.

In practical terms, environmental biogeochemistry can be considered a branch of biogeochemistry, studying specifically the effects humans have on the natural environment, or how the natural processes and cycles are affected by human activity. As such, environmental biogeochemistry will necessarily also study the anthroposphere in detail.

There are many fields of application of environmental biogeochemistry, broadly defined as the theoretical and the practical. Theoretical developments aim to understand the anthroposphere processes and mechanisms, with applications such as modelling of natural systems, eutrophication (over-mineralization) of surface waters, climate change effects on the biosphere, and soil chemistry, to name a few. Practical applications include soil and water acidification remediation, pollution remediation, carbon sequestration, and biogeochemical prospecting for mineral deposits, among others. Here we will consider one such practical application, the remediation of heavily metal-contaminated and acidic soil.

5.4.2   Environmental Biogeochemistry and Remediation

An industrial wasteland near San Francisco Bay (California, USA) had a soil so severely polluted that it was unable to support any vegetation and, by extension, any form of animal life; it was a clearly toxic environment. On average, the soil contained 0.3% arsenic, 0.18% mercury, 1.26% lead, and 0.24% selenium and its pH was between 3.6 and 3.8 [10]. This is unfortunately not an uncommon situation and is observed in the vicinity of virtually any strongly industrialized area or smelter (□ Fig. 5.13). Because of that, scientists have been trying to find practical and effective, but also economical ways to combat both the pollution and the acidity of the soil. This is done by adding some natural material—rocks or organic matter—that is relatively readily available and that can improve the quality of the soil. For instance, adding ground limestone to acidic and polluted soils can result in an increase of pH from 3.5 to near neutral 6.5 [21]. It also has two other beneficial effects: toxic metals can precipitate as carbonates, thus reducing their bioavailability, and Ca (and Mg, if dolomite is used) provide nutrition for plants and, at high levels,

Fig. 5.13 Copper smelter and sulphuric acid plant in Copperhill (Tennessee, USA), as seen in 1939. Massive emissions of sulphur in the air, resulting from processing the sulphide ore, led to severe acidification of the groundwater and the ultimate removal of all soil. However, after the smelter was decommissioned, strong remediation efforts resulted in the restoration of the site to near-natural state [6]

Fig. 5.14 City of Flin Flon (Manitoba/Saskatchewan, Canada) suffered from complete soil loss due to the sulphur emissions from the lead and zinc smelter processing sulphide ores. As the smelter is winding down and is expected to close down in the near future, remediation efforts are under way, consisting of placing crushed limestone on the rocks. These efforts have resulted, in a decade, in the beginning of new soil and biomass production. The limestone provided de-acidification, but also sites for trapping of organic matter and the nucleation of new soil

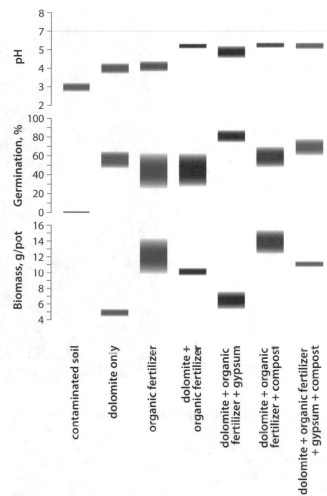

Fig. 5.15 Effects of different soil amendments on soil pH, seed germination, and biomass production of Fawn tall fescue (*Festuca arundinacea* Schreber), which proved to be the most tolerant plant species. Results for untreated contaminated soil are presented, as well. The biomass and soil pH were determined 60 days after seed germination. Data from Lin et al. [10]

protect plants against the toxic metals (Ca and Mg will be consumed preferentially; [1]). This simple but highly effective method is often employed in toxic and acidic soils or in areas where soils were removed, such as the century-old mining area in the towns of Flin Flon (Manitoba/Saskatchewan, Canada; ☐ Fig. 5.14) or Sudbury (Ontario, Canada).

In the particular case of San Francisco, scientists conducted a thorough investigation of what other additives, besides dolomite, can be beneficial to soil remediation, or phytorestoration. In addition to rock material (dolomitic limestone and gypsum), they used organic fertilizer and organic compost (from mushroom production). Samples of contaminated soil were collected,

and then experiments were conducted in a controlled environment, a greenhouse. Several experiments were conducted using various combinations of the additive materials, from just dolomite to all of them together. The soils were planted with seeds from seven plant species to evaluate the most tolerant ones to this soil: Fawn tall fescue (*Festuca arundinacea* Schreber), perennial ryegrass (*Lolium perenne*), harding grass (*Phalaris aquatia* var Perla), California brome (*Bromus carinatus*), Carolina poplar (*Populus canadensis* L.), streaker redtop (*Agrostis gigantean* Roth), Reubens Canada bluegrass (*Poa campressa* L.), and Indian mustard (*Brassica juncea* L.).

The results were evaluated in terms of three main factors: acidity reduction, seed germination (percentage of the seeds that managed to germinate), and shoot biomass production (grams of material for each experimental pot; ☐ Fig. 5.15). The results from the different additive combinations were compared and give a very

◘ **Fig. 5.16** Scanning electron microscope images of water treatment residuals (WTR). **a** typical rough and smooth surfaces of aluminium-rich WTR; **b** typical rough and smooth surfaces of iron-rich WTR; **c**: close-up of the rough surface of aluminium-rich WTR particle; **d** magnified secondary image of the close-up of the rough surface of iron-rich WTR particle. Images from Makris and O'Connor [12], modified

clear idea of the best additives combination, at least for this particular situation.

The results (◘ Fig. 5.15) are very clear: if the soil is left without any treatment, there will be virtually no seed germination, resulting in no biomass being produced at all. Adding just dolomite had limited effects, with pH increasing to 4, germination to 57%, and biomass production to 5 g per pot. The highest germination, at 81%, was achieved when dolomite, gypsum, and organic compost were added to the soil; the highest pH was achieved by several combinations of mineral and organic additives, and the highest biomass production was achieved by adding dolomite, organic compost, and organic fertilizer (◘ Fig. 5.15; [10]). The most resilient plant proved to be Fawn tall fescue (*Festuca arundinacea* Schreber), with 35% germination and 4 g of biomass per experimental pot [10]. The authors of this study were able to provide a relatively simple "recipe" for the *phytorestoration* of similarly contaminated soils, concluding that the optimum commination to effectively reduce soil acidity and ameliorate plant growth consisted in adding 0.9% of dolomite and 1.3% of organic fertilizer. This example illustrates that it is possible, with relatively simple means, to effectively address one of the most pressing environmental concerns, the remediation of severe industrial pollution.

5.4.3 Adsorption and Remediation

There are other ways to address remediation, and geochemistry provides some valuable resources in that respect. We mentioned in the previous section that toxic metals can precipitate as carbonate minerals, provided soil pH is increased, and make them less bioavailable. We also described in ▸ Chap. 4 how elements adsorb on the surface of specific minerals, often clays (illite, kaolinite) or Fe and Mn oxides. In that case, this phenomenon was used for exploration purposes: the element of interest migrated from a deposit at depth and was trapped in the soil to form a surface anomaly. We also discussed the strong dependence of adsorption on the pH of the soil (◘ Fig. 4.6).

The same principles and ideas can be effectively applied to pollution remediation. The general principle is to maximise, by modifying the pH and using additives, the adsorption of toxic metals on specific clay or oxide minerals and thus reduce their bioavailability. In practice, any material that is readily available in relatively large amounts and is affordable can be used as additive, on condition that it achieves higher pH and lower toxicity. So far, quarried crushed limestone, dolomite, or some clay-rich rocks have been used to remediate soil pollution, but these come at a cost and generate their own production-related pollution, making them slightly counter-productive in terms of environmental protection. Scientists have been experimenting with a variety of other materials, such as biochar, produced by pyrolysis of biomass and commonly used for carbon sequestration and for soil health benefits. However, some people have come up with an original and elegant solution: use other types of waste, in particular the water treatment residual (WTR) solid waste.

The majority of drinking-water treatment plants are based on coagulation and filtration. Fe, Al, and

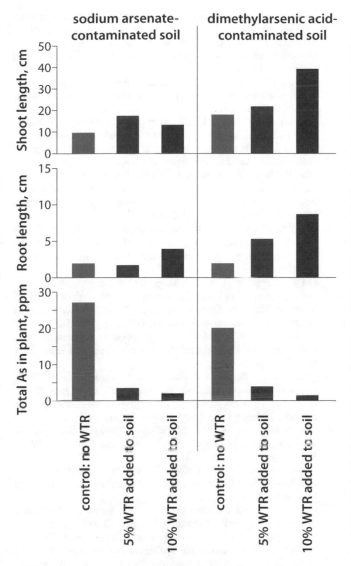

Fig. 5.17 Rice shoot length, root length, and amount of arsenic as function of the amount of Fe-rich WTR added to the soil. The experiments were conducted for two common forms of arsenic contamination in soil (sodium arsenate and dimethylarsenic acid), and in both cases, the results show clear beneficial effects of adding WTR to the soil. From Datta et al. [5], modified

The solid waste produced by the water treatment, a.k.a. water treatment residuals (WTR) is most often disposed of in landfills. It is estimated that approximately 2 tons of WTR are produced each day in the USA alone [16], leading people to suggest that WTR can be a low-cost soil amendment to reduce environmental impacts of various oxyanions, notably phosphorus and arsenic.

The WTR produced during water treatment have widely different elemental compositions and sorption capacities. They have broad particle size distributions, but rarely have particles larger than 5 mm, with most particles falling in the 1 to 10 mm range ([12]; Fig. 5.15). Al-based WTR are typically slightly acidic (pH of 5–7), whereas Fe-based ones are normally alkaline (pH of 7–9). Both types are predominantly amorphous, with little crystalline content. Observations by scanning electron microscope show that particles in both Al-based and Fe-based WTR have irregular shapes and surfaces, from smooth to rough (Fig. 5.15).

Crucially, these WTR characteristics make them very good in amending soils that are naturally poorly sorbing and cannot effectively sequester such common contaminants such as phosphorus and arsenic on their own. Arsenic, in particular, is highly toxic and is often present in the soil in high concentrations due to pesticides, but also due to treated wood found in decks, docks, and playgrounds. Importantly, if contaminants are not immobilised in the soil (typically by sorption), they will be ingested by and accumulated in plants, making them very dangerous for human consumption. However, adding WTR is an effective way of increasing the sorbing qualities of the soil, as a recent study demonstrated [5]. In this study, rice was planted in soil contaminated with 150 ppm of sodium arsenate (Na_3AsO_4) or dimethylarsenic acid ($C_2H_7AsO_2$), two common forms of arsenic in soil. Several factors were studied as function of the amount of WTR added to the soil (5 or 10%), and the results (Fig. 5.17) are quite clear.

In terms of plant growth, the effects of adding WTR are more pronounced for soil contaminated by dimethylarsenic acid, even though they show improved growth for sodium arsenate-contaminated soils, as well (Fig. 5.16). The most dramatic effects are in terms of plant ingestion of arsenic, in particular for sodium arsenate-contaminated soils (Fig. 5.16), demonstrating that arsenic was much more effectively sorbed on the surface of WTR particles and made much less available for plants to ingest.

Overall, these are very encouraging results demonstrating that a previously discarded by-product, the solid residual material produced by water treatment, can be an effective toxicity-reducing additive to contaminated soils. The fundamental principle here is the same as previously discussed: we add a specific material to the soil that will increase its pH and increase adsorption (both in terms of amounts and of binding) of toxic elements making them much less bioavailable.

Ca salts, as well as polymers and activated carbon are added to raw water to remove colloids, colour, sediment, and common contaminants from surface and groundwater that is intended for use as drinking water [12]. The addition of Fe or Al salts in particular leads to the adsorption or precipitation, at basic pH, of common contaminants such as phosphorus and arsenic: the same process is used in soil remediation (e.g., Livesey and Huang [11]). The polymers and the activated carbon, on the other hand, are used to remove different inorganic and organic contaminants (e.g., Maurer and Boller [13]). All in all, this is a well-established technology that is very effective in removal of most common organic and inorganic contaminants from groundwater to produce drinking water.

5.5 Summary

The natural environment has been strongly impacted by human activity, both by our utilization of natural resources and by emitting polluting materials, to the point of causing irreversible modifications to all of Earth's components. Today, the interactions between the anthroposphere and the various components of the Earth are subject of intense studies, with the purpose to understand their complexities, trace and study industrial and agricultural pollution, and provide tools and methods of remediation. In that, environmental geochemistry and environmental biogeochemistry are valuable methodological frameworks.

Tracing pollution using geochemical methods is not dissimilar to mineral exploration, in that we measure elemental concentrations and isotopic values in a variety of surface sampling media (groundwater, soil, vegetation) to discover the source of industrial, agricultural, or other pollution, to quantify its extent in terms of space and concentrations, and to have an idea of the pollutant dispersion mechanisms.

Geochemistry is also very useful in remediation studies, as it provides us with knowledge about the mobility of elements in the surface environment. Relatively simple procedures—e.g., adding crushed carbonates to an acidic and polluted soil—can have immediate beneficial effects of increasing pH of the groundwater and increasing the sorption of toxic metals making them less bioavailable.

❓ Exercises

Q5.1 What is the anthroposphere?

Q5.2 What are the most commonly used geochemical indicators of past climate changes?

Q5.3 How can we trace and study industrial and farming-related pollution?

Q5.4 What is environmental biogeochemistry?

Q5.5 How can we address soil loss due to low pH?

Q5.6 What are the most effective ways to address acidic and toxic soils?

References

1. Adriano DC (2001) Trace elements in terrestrial environments: biogeochemistry, bioavailability, and risks of metals. Springer, New York
2. Alexandre P (2020) Isotopes and the natural environment. Springer, New York
3. Aravena R, Evans ML, Cherry JA (1993) Stable isotopes of oxygen and nitrogen in source identification of nitrate from septic systems. Groundwater 3:180–186
4. Corcoran S (2019) Nitrate contamination and nitrogen isotope compositions in the Bazile groundwater management area—a water quality study. Environmental Studies Undergraduate Student Theses, University of Nebraska-Lincoln, Creighton, NE, p 248
5. Datta R, Sarkar D, Hussein H, Therapong C (2007) Remediation of arsenical pesticide applied soils using water treatment residuals: preliminary greenhouse results. In: Sarkar D, Datta R, Hannigan R (eds) Developments in environmental science, vol 5. Elsevier, Oxford, pp 543–560
6. Dixon C (2007) Up from the mines in Tennessee. New York Times. ▶ www.nytimes.com/2007/12/14/travel/escapes/14copperhill.html. Retrieved 23 Mar 2020
7. Hollabaugh CL (2007) Modification of Goldschmidt's geochemical classification of the elements to include arsenic, mercury, and lead as biophile elements. In: Sarkar D, Datta R, Hannigan R (eds) Developments in environmental science, vol 5. Elsevier
8. Hou X, Parent M, Savard MM, Tasse N, Begin C, Marion J (2006) Lead concentrations and isotope ratios in the exchangeable fraction: tracing soil contamination near a copper smelter. Geochem Exploration Environ Anal 6:229–236
9. Kuhn A, Heckelei T (2010) Anthroposphere. In: Speth P, Christoph M, Diekkrüger B (eds) Impacts of global change on the hydrological cycle in West and Northwest Africa. Springer, Berlin, pp 282–341
10. Lin Z-Q, Hussein H, Ye ZH, Terry N (2007) Phytorestoration of metal-contaminated industrial wasteland: a greenhouse feasibility study. In: Sarkar D, Datta R, Hannigan R (eds) Developments in environmental science, vol 5. Elsevier, Oxford, pp 487–502
11. Livesey NT, Huang PM (1981) Adsorption of arsenate by soils and its relationship to selected chemical properties and anions. Soil Sci 131:88–94
12. Makris KC, O'Connor GA (2007) Beneficial utilization of drinking-water treatment residuals as contaminant-mitigating agents. In: Sarkar D, Datta R, Hannigan R (eds) Developments in environmental science, vol 5. Elsevier, Oxford, pp 609–639
13. Maurer M, Boller M (1999) Modeling of P precipitation in wastewater treatment plants with enhanced biological P removal. Water Sci Technol 39:147–163
14. Peters SE, Carlson AE, Kelly DC, Gingerich PD (2010) Large-scale glaciation and deglaciation of Antarctica during the Late Eocene. Geology 38:723–726
15. Petit JR, Jouzel J, Raynaud D, Barkov NI, Barnola JM, Basile I, Bender M, Chappellaz J, Davis M, Delaygue G, Delmotte M, Kotlyakov VM, Legrand M, Lipenkov VY, Lorius C, Pepin L, Ritz C, Saltzman E, Stievenard M (1999) Climate and atmospheric history of the past 420,000 years from the Vostok ice core, Antarctica. Nature 399:429–436
16. Prakash P, Sengupta AK (2003) Selective coagulant recovery from water treatment plant residuals using Donnan membrane process. Environ Sci Technol 37:4468–4474
17. Robertson WD, Cherry JA, Sudicky JA (1991) Ground-water contamination from two small septic systems on sand aquifers. Groundwater 29:82–92
18. Stips A, Macias D, Coughlan C, Garcia-Gorriz E, Liang XS (2016) On the causal structure between CO2 and global temperature. Sci Rep 6:21691
19. Veizer J, Ala D, Azmy K, Bruckschen P, Buhl D, Bruhn F, Carden GAF, Diener A, Ebneth S, Godderis Y, Jasper T, Korte C, Pawellek F, Podlaha OG, and Strauss (1999) 87Sr86Sr, 13C and 18O evolution of Phanerozoic seawater. Chem Geol 161:59–88
20. Zachos J, Pagani M, Sloan L, Thomas E, Billups K (2001) Trends, rhythms, and aberrations in global climate 65 Ma to present. Science 292:686–693
21. Winterhalder K (1996) Environmental degradation and rehabilitation of the landscape around Sudbury, a major mining and smelting area. Environ Rev 4:185–224

Further Reading

22. **Trace Elements in Terrestrial Environments: Biogeochemistry, Bioavailability, and Risks of Metals**, by DC Adriano (2001). Springer, ISBN 978-0-387-98678-4.

23. **Impacts of Global Change on the Hydrological Cycle in West and Northwest Africa**, by P Speth, M Christoph, and B Diekkrüger (2010). Springer, ISBN 978-3-642-12956-8.

24. **Concepts and Applications in Environmental Geochemistry**, edited by D Sarkar, R Datta, and R Hannigan (2007). Elsevier, ISBN 978-0-08-046522-7.

25. **Principles of Environmental Geochemistry**, by GN Eby (2004) Waveland Press, Long Grove. ISBN 978-1-4786-3164-4.

26. **Handbook of Environmental Isotope Geochemistry**, edited by M Baskaran (2011) Springer Science & Business Media, ISBN 978-3-642-10637-8.

Supplementary Information

Epilogue – 102

Glossary – 105

Index – 111

© Springer Nature Switzerland AG 2021
P. Alexandre, *Practical Geochemistry*,
Springer Textbooks in Earth Sciences, Geography and Environment,
https://doi.org/10.1007/978-3-030-72453-5

Epilogue

We have arrived at the end of this book, a brief and straightforward introduction to geochemistry. Hopefully you have obtained a clear and solid understanding of the fundamentals and basic principles. It is now time to take a moment and consider the future of geochemistry, and of your own learning.

The Future of Geochemistry

The past and the future of geochemistry have something in common: they are both heavily dependent on technology. New analytical developments have driven novel geochemical applications in the past and will continue to do so in the future. Of course, it would be simplistic to say that the science and understanding of geochemistry will be dependent *uniquely* on technological advancement: scientists are those who will make the best use of technology, and without their expertise, inquisitiveness, imagination, and critical thinking, all technology would be futile. It is also scientists who are the main catalyst for technological developments in geochemistry or other natural sciences fields. However, it is analytical developments that are helping geochemistry advance the fastest, so let us consider a few examples.

Portability

Mineral exploration has always demanded faster turnaround times (the times between sample collection and reporting of results). The ideal scenario for the exploration industry is when the data are obtained directly in the field, removing the whole process of collection, preparation, and transportation of samples, and reducing the wait time for results to virtually zero (and also removing the associated costs). Because of this demand, several technological advancements directly target portability: or the development of analytical techniques that can be used in the field. Two methods already exist, the portable X-ray fluorescence (p-XRF), which we discussed earlier, in Chapter 2, and the Portable Infrared Mineral Analyzer (PIMA). PIMA is much more of a mineralogical identification tool, capable of quantifying the clay minerals present in a sample, which is very useful as most mineral deposits have a cortege of diagnostic hydrothermal clay alterations. Both techniques give results in real time, albeit with analytical uncertainty that is still well below that of laboratory-obtained data.

However, geochemical analytical advance is seeking to further miniaturize and mobilize any analytical technique possible. We can foresee a time when sample preparation and dissolution, and analysis by quadrupole mass spectrometers or by spectroscopy methods (absorption or emission) will be conducted in the field. Microbeam methods, such as the scanning electron microscope, might also find their way in the exploration camp some day. The most likely possibility though is that completely new techniques, possibly based on existing analytical methods, will be developed specifically for use in the field.

Ease of Use

Thankfully, long gone are the times when analytical installations were slow, unstable, and unreliable. An area of constant progress in analytical geochemistry is the stability and the ease of use of the specific tools. One example is the quadrupole mass spectrometer (and the associated inductively coupled plasma), which is now a mature (i.e., very well developed) technology. This is the perfect example of a new generation of ready-to-use, or plug-and-play, machines. The scientist is no longer required to build the accessories (e.g., gas extraction and purification lines or even the clean laboratory dissolution and element separation techniques): they now come off-the-shelf, ready to use and integrate with other accessories and with the mass spectrometer.

One result of this development is that the labour involved in sample preparation and dissolution has been drastically reduced. The graduate student in geochemistry used to spend the vast majority of their time in the laboratory, which was tedious, uncomfortable, and sometimes unpleasant; it was considered the normal thing to do. Not so any more, when automation has become much more prevalent.

Automation

One natural consequence of analytical progress is the further automation of any step of the sample preparation and analysis process. This also includes, amazingly, the whole clean laboratory portion of sample preparation, which has always been a highly specialized and difficult job prone to errors. Virtually any task can be automated, and it is often very advantageous to do so for the following two major reasons: (1) reducing the possibility of human error, such as sample cross-contamination, and increasing the reliability of the data, and (2) processing a larger number of samples and doing so faster. If we can have more reliable data obtained more quickly (and cheaper), who would complain? Certainly not the exploration industry, for which the turn-around time is always a significant issue (as mentioned earlier); and certainly not the research scientist who can base their findings on more numerous and more reliable data.

Lower Detection Limits

As automation is becoming more and more common and as new types of detectors are developed (and old types improved), the detection limits are constantly decreasing. Regardless of the analytical method, concentrations in the lower parts per billion (ppb) are now routinely and reliably measured for most elements; parts per trillion (ppt) are commonly measured in any self-respecting laboratory, and even detection limits as low as parts per quadrillion (ppq) are not unheard-of and are becoming less exotic. This is amazing and the effects on both the exploration industry and academic research are profound. The exploration geologist will be able to detect weaker and weaker geochemical anomalies, produced by deeper and deeper deposits, while the scientist will be able to utilize and study ultra-trace elements, which had not been possible before.

Your Future Learning

Recommended Books

Some of you might consider delving further into the topic of geochemistry and maybe even making a career of it (applied or academic); others might want to learn about a specific aspect of geochemistry and use it in your work. For those of you who are interested, here are a few suggested books and a set of practical exercises that will help you apply your new knowledge. There are plenty of other books available, so the following is only a selection of appropriate books to advance further in a specific area of geochemistry.

Isotope Geochemistry and Geochronology
- Isotope Geochemistry, W.M. White, 2015 (Wiley-Blackwell, ISBN 987-0-470-65670-9). A solid and straightforward text.
- Isotope Geology, C. A. Allegre, 2008 (Cambridge University Press, ISBN 978-0-511-45524-7). Brilliant, but not for the faint-hearted: it is a fiercely intense and highbrow text, only for the fairly advanced student.

- Isotopes, Principles and Applications, G. Faure and T. M. Mensing, Wiley, 2018, ISBN 978-81-2653837-9. This is the definite and complete treatise on isotopes, complete and up-to-date. At 900 pages, a bit too long perhaps.
- Isotopes and the Natural Environment, P. Alexandre, 2020 (Springer, ISBN 978-3-030-33651-6). A very good introduction to isotopes applied in variety of natural sciences.
- Stable Isotope Geochemistry, J. Hoefs, 2013 (Springer, ISBN 978-354-061126-4). A standard text that has stood the test of time.
- Radiogenic Isotope Geology, A. P. Dickin, 1995 (Cambridge University Press, ISBN 978-052-143151-4). Solid standard text.
- Geochronology and Thermochronology by the $^{40}Ar/^{39}Ar$ Method, I. McDougall and T. M. Harrison, 1999 (Oxford University Press, ISBN 978-0-195-10920-7). The bible of $^{40}Ar/^{39}Ar$ geochronology.

Exploration Geochemistry
- Introduction to Exploration Geochemistry, 2nd edition, A. A. Levinson, 1980 (Applied Publishing Ltd., ISBN 0-915834-04-9). Solid but dated.
- Applied Geochemistry: Advances in Mineral Exploration, A. S. Macheyeki, D. P. Kafumu, X. Li, F. Yuan, 2010 (Elsevier, ISBN 978-0-128-19495-9). Good, with a few case studies.
- Exploration Geochemistry, A. S. Joyce, 1976 (Techsearch Incorporated, 978-0-909-38604-7). Clear but dated.
- Biogeochemistry in Mineral Exploration, C. Dunn, 2007 (Elsevier Science, ISBN 978-0-444-53074-5). Brilliant, thorough, up-to-date.

Environmental Geochemistry
- Principles of Environmental Geochemistry, G. N. Eby, 2004 (Waveland Press, ISBN 978-1-4786-3164-4). Well written, good coverage of the fundamentals.
- Concepts and Applications in Environmental Geochemistry, edited by D. Sarkar, R. Datta, and R. Hannigan, 2007 (Elsevier, ISBN 978-0-08-046522-7). An excellent compilation of up-to-date papers.
- Trace Elements in Terrestrial Environments: Biogeochemistry, Bioavailability, and Risks of Metals, D. C. Adriano, 2001 (Springer, ISBN 978-0-387-98678-4). Excellent book.
- Handbook of Environmental Isotope Geochemistry, edited by M. Baskaran, 2011 (Springer, ISBN 978-3642106361). Very good and up-to-date.
- Isotope Tracers in Catchment Hydrology, C. Kendall, J. J. McDonnell, J. J. McDonnell, 1998

(Elsevier, 978-044-481546-0). A bit dated, but still a good source.

Analytical Geochemistry and Data Treatment

- Modern Analytical Geochemistry, edited by R. Gill, 1997 (Longman, ISBN 0-582-09944-7). An excellent book, to be used extensively and without moderation, even though it's a bit dated.
- Isotopic Analysis: Fundamentals and Applications Using ICP-MS, edited by F. Vanhaecke and P. Degryse, 2012 (Willey, ISBN 978-3527328963). Very useful.
- Using Geochemical Data: Evaluation, Presentation, Interpretation, H. Rollinson, 1993 (Pearson Education, ISBN 0-582-06701-4). A useful book, with plenty of relevant and practical information.

Practical Exercises

These exercises are designed to give you the opportunity to practice your new knowledge and understanding of geochemistry. They are open-ended and don't have a right or wrong answer: you are free to develop them as you wish. When you work on them, try to consider all possible aspects and implications; try also to be detailed, specific, and convincing.

Lithogeochemistry

You are a PhD student and have been given the rare opportunity—as a member of a large international team—to study some igneous and metamorphic rocks in northern Greenland that have only recently been exposed due to ice melting and have never been studied before. Using geochemical methods, you are going to help understand the history and evolution of these rocks, including when, where, and by what processes and under what conditions they formed and evolved. Due to substantial government investment, you have the luxury to use any methods you wish, but you have to develop a plan and submit it for the approval of the government. In the plan (less than five pages long), you must describe in detail what geochemical approaches you are going to use, and what you expect to learn from each of them.

Exploration Geochemistry

You have been hired to the mineral exploration team of a major mining company in Chile, and your job is to discover new porphyry copper deposits in the north of the country. The area is mountainous (high plateau), very dry and fairly cold, with little soil cover. The expected deposits are situated close to the surface (200–400 m), but are sometimes deeper. The main ore-related minerals are pyrite, chalcopyrite, bornite, pyrrhotite, molybdenite, and arsenopyrite, disseminated in strongly altered granitoids.

You task is to design and implement a geochemical exploration program lasting two years. You are in charge of all major decisions regarding planning and conducting the program, interpreting the results, and defining a target area for further detailed exploration and possibly drilling. Design the program considering at minimum the following questions:

- What are the stages of the program?
- What are the sampling media you are going to use?
- How many personnel are you going to need and for how long?
- What is the sampling density; what is the sampling pattern?
- What field technologies are you going to use? What is the benefit of using them?
- What analytical methods are you going to use? Describe in detail how these methods work, including theoretical background and methodological considerations.
- What elements are you going to analyze for?
- What are the data interpretation methods you are going to use?
- What are the environmental, economic, and societal impacts of your exploration campaign?

Write your project (not to exceed five pages) in a convincing manner, so that it can be approved by the upper management: give detailed and specific information about the methodology and the logistics of the campaign.

Industrial Pollution

A city of 150,000 people in North America had a brief and intense industrial production period ending a century ago, producing steam engines, farming equipment, and other heavy machinery. That left significant portions of the city's soil heavily polluted by heavy metals, in particular chromium, vanadium, and nickel. Now the city wants to develop these areas for residential and commercial use, but needs to assess (1) where the most polluted areas are, and (2) how to best remediate the pollution.

You, as an independent environmental geochemistry consultant, have been contracted to develop a plan to answer those two questions. Using your knowledge and experience, develop a specific and detailed plan, considering all the possible aspects of the task, such as sampling media, analytical methods, logistics, data interpretation, remediation techniques, their cost and effectiveness, time frame, and any other relevant factor. Propose two alternative plans at different cost level ("cheap-and-dirty" and "all the bells-and-whistles") for the city administrators to choose.

Glossary

A

Absorption spectrum: The characteristic spectrum that a chemical element absorbing electromagnetic radiation produces when it is observed with a spectroscope. Black lines or bands appear where energy has been removed from the continuous spectrum by an absorbing element.

Acidification: The decrease of pH of soil or other surface medium.

Actinides: One of the 15 metallic chemical elements with atomic numbers from 89 to 103, actinium through lawrencium.

Adsorption: The adherence or attachment of ions or molecules in solution to the surface of minerals with which they are in contact.

Albedo: The ratio of the amount of electromagnetic energy reflected by a surface to the amount of energy incident upon it. Usually albedo refers to the reflectivity of solar energy.

Alkali metal: A monovalent element, one of lithium, sodium, potassium, rubidium, and caesium, situated in group 1 of the periodic table.

Alkali-earth metal: One of the six chemical elements in group 2 of the periodic table: beryllium, magnesium, calcium, strontium, barium, and radium.

Alteration: Any change of the mineralogy of a rock brought caused by physical or chemical reactions, in particular by hydrothermal solutions. Also, the changes in the chemical or mineralogical composition of a rock by weathering.

Anion: Negatively charged ion.

Anomaly: A departure from the typical, normal, or expected values of some chemical or physics property.

Anthroposphere: That part of the environment that is made or modified by humans for use in human activities and as human habitats.

Aquifer: A body of rock containing saturated permeable material and able to conduct groundwater.

Aquitard: A body of rock with low permeability preventing it from conducting groundwater.

Argon-argon geochronology: An absolute (isotopic) geochronological method, based on the disintegration of ^{40}K to ^{40}Ar. It is a development of the K–Ar method, involving the production of ^{39}Ar by irradiation of the sample, using it as a proxy for the amount of K.

Atmophile element: In Goldschmidt's classification, a chemical element (O, N, Ar, N) that is cumulated preferentially in the atmosphere.

Atmosphere: The mixture of gases that surrounds the Earth. It consists of 78% nitrogen, 21% oxygen, 0.9% argon, 0.04% carbon dioxide, and trace amounts of helium, krypton, neon, and xenon.

Atomic mass: The mass of an atom, almost entirely made of the masses of protons and neutrons, and measured in daltons (Da, 1/12 of the mass of ^{12}C atom at rest).

Atomic number: The unique number for each chemical element, corresponding to the number of protons in its nucleus.

Atomic spectroscopy: An analytical technique whereby a sample is vaporized and its atoms are identified and quantified by the electromagnetic radiation they absorb.

B

Basalt: A very common mafic extrusive igneous rock formed from the rapid cooling, near or at the surface of the Earth, of Mg- and Fe-rich lava.

Big Bang: A cosmological model of the universe from the earliest known periods through its subsequent large-scale evolution, describing how it expanded from an initial state of infinitely high density and high temperature and offering a comprehensive explanation for many observed phenomena.

Bioavailability: The measure of how much of a chemical element in the environment may enter into living organisms. It is a limiting factor in the production of crops and in the removal of toxic substances from the environment.

Biogenic dispersion: Dispersion of elements or particles from rock into the surface environment by living organisms.

Biogeochemistry: A branch of geochemistry dealing with the interactions between rocks and the biosphere.

Biomass: The amount of living material in a specific area, measured in terms of weight or volume of organisms per unit area or per unit volume of the environment.

Biophile elements: Those elements that are the most typical in organisms and organic matter or that are concentrated in living organisms.

Biosphere: All living organisms of the Earth or the area occupied by them. It overlaps with the lithosphere, hydrosphere, and atmosphere.

Bulk Silicate Earth (BSE): The original chemical composition of the silicate part of the Earth after its accretion and separation of a Fe-rich core but prior to differentiation of the primitive crust.

C

Calc-alkaline: A series of igneous rocks typically occurring at convergent margins and showing a trend of iron enrichment with increasing silica content.

Carbon sequestration: The long-term removal, capture, or sequestration of carbon dioxide from the atmosphere to slow or reverse atmospheric CO_2 pollution and to mitigate or reverse global warming.

Cation: A positively charged ion.

Chalcopile elements: Those elements (Ag, As, Bi, Cd, Cu, Ga, Ge, Hg, In, Pb, S, Sb, Se, Sn, Te, Tl, and Zn) that readily combine with sulphur.

Chart of the isotopes: A plot of the number of neutrons vs. number of protons in a nucleus, on which any known isotope can be found.

Chemical dispersion: The mobilization of chemical elements from a rock or a deposit by chemical processes, typically due to hydrothermal alteration or weathering.

Chondrite: A meteorite containing chondrules and a primitive bulk chemical composition, often taken as proxy for the bulk chemical composition of the solar system.

Climate change: The set of climate changes due to the rising mean average atmosphere temperatures.

Closure temperature: In geochronology, the temperature at which the loss (by diffusion) of a radiogenic daughter isotope will be negligible relative to its production by radioactive disintegration.

Colluvium: A general term applied to any loose, heterogeneous, and incoherent mass of soil and rock fragments deposited by rainwash, sheet wash, or soil creep, usually collecting at the base of slopes or hillsides.

Commodity: In mineral exploration, the element or rock of interest which represents the main economic interest.

Common lead: In geochronology, lead that is not produced by radioactive disintegration of U or Th.

Common lead correction: The subtraction of the amount of common lead incorporated in a mineral when it formed initially.

Compatible element: An element that, during any igneous process, enters preferentially the solid phase co-existing with a liquid phase.

Concordia: In the concordia diagram, a curved line on which the $^{207}Pb/^{235}U$ and the $^{206}Pb/^{238}U$ ages are the same.

Concordia diagram: A diagram on which $^{207}Pb/^{235}U$ ratios are plotted against the $^{206}Pb/^{238}U$ ratios for the same sample.

Contamination: The introduction of foreign substance in a sample, leading to the modification of its properties and chemical composition.

Continental collision: In plate tectonics, the collision between two converging sections of continental crust.

Correlation: In statistics, any statistical relationship, causal or not, between two random variables or bivariate data.

Correlation coefficient: The strength of the correlation between two random variables.

Correlation table or matrix: A table summarizing the correlation coefficients between several variables.

cosmogenic isotope: Isotope produced in the upper reaches of the atmosphere by cosmic rays radiation.

D

Diagenesis: All chemical and physical changes in minerals after their initial accumulation as sediments, due to the increasing pressure and temperature during burial. Changes include addition and removal of material, dissolution, recrystallization or replacement, and phase changes.

Differentiation: All processes by which multiple rock types form from a single magma source.

Discordia: A straight line in the concordia diagram, intersecting the concordia line in two places, the upper and the lower intercept. These are most commonly interpreted as the age of initial crystallization and the age of partial radiogenic Pb loss.

Discrimination diagram: Any diagram that is used to differentiate two rocks on the basis of their chemical composition.

Dissolution: The process of dissolving a solid into a homogenous solution.

E

Earth's mantle: The zone of the Earth below the crust and above the core, which is divided into the upper mantle and the lower mantle, with a transition zone between.

Eh: Oxidation potential, measured as the potential of a half-cell, measured against the standard hydrogen half-cell.

Electron: A subatomic elementary particle of the lepton particle family with one elementary negative charge and a mass approximately 1/1836 that of the proton.

Electron microprobe: An microbeam analytical technique consisting of bombarding the surface of a sample with an electron beam and detecting the characteristic X-rays emitted by the elements in the sample, providing quantitative chemical analysis of the sample.

Electronegativity: The tendency of an atom to attract a shared pair of electrons (or electron density) towards itself.

Element compatibility: The likelihood of an element to preferentially fractionate into the solid (crystalline) phase during any igneous process.

Element mobility: The extent to which an element can be mobilized under a set of physical and chemical conditions, for instance in the surface environment in the presence of free water and oxygen.

Element substitution: The replacement of an element by another in the crystalline structure of a mineral.

Emission spectrum: In spectroscopy, the characteristic spectrum emitted by an element or a group of elements in a dissolved sample, while being heated or bombarded with photons, ions, or electrons.

Enriched Mantle: A part of the Earth's mantle in which the concentrations of relatively incompatible elements have been increased.

Environmental geochemistry: The application of geochemical theories, principles, methods, and techniques to (1) the study of the natural environment and of our interactions with it, and (2) the remediation of pollution.

Europium anomaly: The deviation from the expected value of Eu in the Rare Earths Spectrum, due to its incorporation in plagioclase (replacing Ca) during magma crystallization. Alternatively due to the input of high-temperature hydrothermal fluids into waters in which chemical sediments precipitate.

F

Faraday cage: In analytical geochemistry, a detector consisting of a small metallic box and a wire leading to a resistor (typically 10^{11} or 10^{12}). Any ion hitting the cage will dislodge one electron which is them multiplied by the resistor to a detectable electric current proportional to the number of ions hitting the detector.

Felsic: Any igneous rock having abundant light-coloured minerals in it. Chemically, any igneous rock with more than approximately 63 wt% SiO_2.

Foraminifera: Predominantly marine members of a class of amoeboid protists characterized by streaming granular ectoplasm (for catching food and other uses) and an external shell of diverse forms and materials.

G

Gangue: The valueless rock or minerals accompanying the ore minerals in an ore deposit. It is separated from the ore minerals during concentration.

Geochemical classification: Any geochemical method used to categorize rocks based on their chemical composition.

Geochemical exploration: Mineral exploration using predominantly geochemical methods.

Geochemistry: The study of the processes affecting the Earth and its various components using their chemical composition.

Geochronology: The study of the ages affecting a rock, typically its initial formation.

Geophysics: The study of the physical properties of the Earth.

Geosphere: The part of the Earth consisting of rocks.

Geostatistics: Statistics applied to geology, characterized by the use of spatial information for each sample.

Geotectonics: A branch of geology dealing with the architecture of the outer part of the Earth, the regional assembling of structural or deformational features, a study of their mutual relations, origin, and historical evolution.

Granite: A felsic intrusive igneous rock, with SiO_2, typically in excess of approximately 69 wt%.

Greenhouse effect: The heating of the atmosphere by the absorption of radiative energy from the sun due to the presence of gases that have higher absorbency.

H

Heavy isotopes: The isotopes with masses typically above 100.

Histogram: A visual representation of the distribution of numerical or categorical data, constructed by defining a series of bins (ranges) of values and counting the number of observations in each bin.

Hydrography: The branch of science dealing with the measurement and description of the physical features of oceans, seas, coastal areas, lakes and rivers, as well as with the prediction of their change over time.

Hydrosphere: The totality of all bodies containing water on Earth.

Hydrothermal: Related to fluids circulating within rocks at depth.

I

ICP-MS: An analytical technique where the sample (aerosol from solution or particles produced by laser ablation) is ionized by a plasma torch (ICP) and introduced and analyzed in a mass spectrometer.

Igneous rock: A rock that solidified from molten or partly molten material, i.e., from a magma.

Incompatible element: An element that, during any igneous process, fractionates preferentially the liquid phase co-existing with a solid phase.

Inductively-Coupled Plasma (ICP): A device where particles are heated in excess of 10,000 °C by inducing them to change orientation at radio frequency by changing the orientation of an electromagnetic field.

Ionic charge: The charge of an ion, determined by the number of electrons in its outermost, valance electron shell.

Ionic potential: The ratio of electric charge to the radius of an ion.

Island arc: A generally curved belt of volcanoes above an ocean-ocean subduction zone.

Island Arc Basalts (IAB): Basalts preferentially found in an island arc.

Isochron: In geochronology, a line of equal ages for the minerals or samples forming it. The slope of the line is proportional to the gae.

Isocon: A line connecting points of equal geochemical concentration.

Isopleth: In general terms, a line or surface on which some numerical function has a constant value.

Isotope: A variety of an element, different from other only by the number of neutrons in its nucleus, but having the same number of protons and therefore the same chemical characteristics but a different mass.

Isotope equilibrium: The state at which an isotope exchange reaction can go either way and isotope fractionation is function only on the conditions of the reaction.

Isotope fractionation: The preferential enrichment of one isotope of an element over another in a compound during a reaction, due to slight variations in their physical and chemical properties. It can be proportional to differences in their masses (mass-dependent fractionation) or not.

Isotopic composition: The proportion of the different isotopes of an element in a rock or mineral.

K

Kriging: A geostatistical method to mostly calculate grades and tonnages of ore reserves from spatially oriented sampling data and relying on the knowledge of the extent and distance of inter-dependence of the observed values.

L

Lanthanides: The 15 chemical elements form La (atomic number 57) though to lutetium (71), sharing common chemical characteristics and predictable behaviour.

Laser ablation: In analytical geochemistry, the bombardment of the surface of a sample by a laser beam to ablate it (extract particles) to be analyzed.

Lead pollution: Any industrial pollution leading to a significant and dangerous increase of lead in the surficial environment.

Light isotopes: Typically, the isotopes of the first 16 element, up to sulphur.

Line of stability: In the chart of isotopes (or chart of nuclides), the line on which stable isotopes are situated. Any isotope not situated on it is radioactive and disintegrates, directly or indirectly, to a stable isotope.

Linear interpolation: A geostatistical method of predicting the value of an location between two points with known values, and assuming that the variation between the two known values follows a linear distribution.

Lithogeochemistry: The study of the chemical composition of whole rocks.

Lithophile element: A chemical element that is concentrated in silicate minerals and therefore tends to concentrate in the Earth's silicate crust.

M

Mafic: An igneous rock composed chiefly of one or more ferromagnesian, dark-coloured minerals in its mode. Its SiO_2 contents is typically between 45 and 52 wt%.

Major elements: The main constituents of a rock (Si, Al, Mg, Fe, Ca, Na, K, and Ti), likely to be present in concentrations above 0.1 wt%.

Mass spectrometer: An analytical instrument designed to separate isotopes by their mass, as they are accelerated and deviated from the straight flight path in a magnetic field.

Mean: In statistics, a measure of the central tendency, calculated as the sum of all values divided by the number of observations.

Mechanical dispersion: The mobilization of fragments of rock by erosion.

Median: In statistics, a measure of the central tendency, defined as the middle value when all values are arranged in increasing order.

Metaluminous: An igneous rock in which the molecular proportion of aluminium oxide is greater than that of sodium and potassium oxides combined but generally less than of sodium, potassium, and calcium oxides combined.

Metamorphic rock: A rock derived from pre-existing rocks by mineralogical, chemical, or structural changes in response to changes mostly in temperature and pressure, generally at depth in the Earth's crust

Meteoric water: Any water (e.g., rain, snow) precipitating on the surface of the Earth.

Microthermometry: The method of estimating pressure, temperature, and chemical conditions of fluid inclusions trapping.

Mid-Ocean Ridge Basalt (MORB): Basalt erupted at a sea-floor spreading centre; by far the most common ocean crust basalt.

Milankovitch cycles: Predictable fluctuations in the seasonal and geographic distribution of insolation, determined by variations of the Earth's orbital, namely eccentricity, tilt of rotational axis, and longitude of perihelion. The respective periods are on the order of 100,000 years, 41,000 years, and 23,000 years.

Mineral exploration: Collectively, all methods (e.g., geophysics, geochemistry) and activates (e.g., prospecting, drilling) employed to discover mineral deposits.

Mississippi-Valley-type (MVT) deposits: Strata-bound lead or zinc deposits in carbonate rocks, often with associated fluorite and barite. They typically have simple mineralogy, occur as veins and replacement bodies, are at moderate to shallow depths, show little post-ore deformation, and are marginal to sedimentary basins.

Mixing calculation: Any calculation to find the chemical or isotopic composition of a mixture knowing the compositions of the two (or more) sources and their relative proportions.

Mode: In statistics, the most commonly observed value, visualized by the highest point in a histogram.

N

Neutron: A subatomic particle, part of the nuclei of atoms, with no net electric charge and a mass slightly greater than that of a proton.

Noble gases: A group of six chemical elements (He, Ne, Ar, Kr, Xe, and Rn) making up group 18 of the periodic table, with similar properties. They are all odourless, colourless, monatomic gases. Under normal geological conditions, they are highly unreactive.

Non-metals: 17 chemical elements that lack the characteristics of a metal. They tend to have a relatively low melting point, boiling point, and density.

Normative mineralogy: The theoretical or calculated mineralogical composition of a rock.

Glossary

Nucleosynthesis: Collectively, the processes that lead to the formation of all chemical elements.

Nugget effect: In geostatistics, the residual variance at zero lag distance, reflecting a variation among replicate measurements and is visualized by a semivariogram whose value is greater than zero at the origin.

O

Ocean Island Basalt (OIB): Basalts formed in intraplate oceanic crust settings, originating from hot spots and chemically distinct from other types of basalts.

Orthometamorphic rock: Metamorphic rock derived from an igneous protolith.

Other metals: Metallic elements (at least Ga, In, Sn, Ta, Pb, and Bi) located between the transition metals and the metalloids in the periodic table. They tend to be soft or brittle, have poor mechanical strength, and have melting points lower than those of the transition metals.

P

Paleotemperatures: Temperatures, typically of the Earth surface environment, in the past.

Parametamorphic rock: Metamorphic rock derived from a sedimentary protolith.

Partial dissolution: In analytical geochemistry, any method of dissolution that purposefully does not dissolve the entirety of the sample.

Partial leaching: Similarly to partial dissolution, mobilization, by a weak acid, or only those elements that are adsorbed on the surface of a mineral.

Pathfinder element: Any element that may serve to indicate the proximity of a mineral deposit.

Pearce Element Ratios: Molar concentration ratio where the denominator is a conserved element; a useful method of quantifying chemical changes of a rock during igneous or hydrothermal processes.

Pearson correlation coefficient: A common statistical measure of the linear correlation between two variables, calculated as the covariance of the two variables divided by the product of their standard deviations.

Peralkaline: An igneous rock in which the molecular proportion of aluminium oxide is less than that of sodium and potassium oxides combined.

Peraluminous: An igneous rock in which the molecular proportion of aluminium oxide is greater than that of sodium and potassium oxides combined.

Percentile: In statistics, a measure indicating the value below which a given percentage of observations in a group of observations falls.

pH: A measure of how acidic (under 7) or basic (above 7) a water-based solution is, defined as the logarithm of the reciprocal of the hydrogen ion activity in a solution.

Phytorestoration: The restoration of any polluted or contaminated soil to its near-natural state, using biological methods.

Porphyry copper deposits: Large low-grade Cu–Mo deposit containing disseminated chalcopyrite and other sulphide minerals, hosted by intermediate to felsic hypabyssal porphyritic intrusive rocks.

Portable Infrared Mineral Analyzer (PIMA): A hand-held mineral exploration tool, used to identify minerals containing water.

Precambrian: Any rocks older than the Cambrian period of the chronostratigraphic scale. The Precambrian period includes the Archean and Proterozoic eons and represents 90% of geologic time.

Primordial nucleosynthesis: The formation of chemical elements during the Big Bang.

Principal Component Analysis (PCA): A geostatistical method, based on correlations, to identify groups of variables with common behaviour.

Product-Moment correlation coefficient: A common statistical measure of nonlinear interdependence of two variables.

Proton: A subatomic particle, part of the atomic nuclei, with a positive electric charge of +1 and a mass slightly less than that of a neutron.

Q

Quadrupole mass spectrometer: In analytical geochemistry, an instrument designed to separate isotopes by accelerating them and deviating them from the straight flight path by an electrostatic field created by four rods to which opposing voltages are applied.

Quartile: In statistics, a measure of location dividing the observed values into four equal parts.

R

Radioactive disintegration (or decay): The process of radioactive isotopes becoming stable by emitting particles and energy.

Radioactive isotope: An isotope that is not stable and constantly disintegrates to become stable.

Radiogenic isotope: An isotope produced by radiogenic disintegration.

Rare Earth Elements (REE): The 15 chemical elements in the periodic table (La through to Lu), also called the lanthanides, which have similar chemical properties.

REE pattern: The pattern formed when the concentrations of REE in an igneous rock are normalized by that of the Bulk Silicate Earth (approximated by a C1 chondrite). The pattern reflects the processes that affected the rock and its degree of differentiation.

Remediation: Any activity aiming to restore a polluted environment to it natural state.

Representativity: The estimation of how representative samples are of the rock they are taken from.

S

Secondary Ion Mass Spectrometry (SIMS): An analytical technique consisting of a beam of ions bombarding the surface of a sample and causing secondary ions to be extracted from it. These are then accelerated and analyzed by a mass spectrometer.

Sampling: In geology, the full set of activities resulting in a representative sample of the relevant medium being collected.

Sampling scheme: The spatial organization of sampling sites in the study area.

Secondary Electron Multiplier (SEM): In analytical geochemistry, a detector consisting of a conversion diode and a series of multiplying plates, producing electrical voltage proportional to the number of ions reaching the detector.

Semi-metals: Also called metalloids, the six chemical elements (normally boron, silicon, germanium, arsenic, antimony, and tellurium) with properties in between, or a mixture of, those of metals and nonmetals.

Siderophile element: Chemical elements (mostly transition metals: ruthenium, rhodium, palladium, rhenium, osmium, iridium, platinum, gold, cobalt, and nickel) that dissolve readily in iron either as solid solutions or in the molten state, and therefore tend to sink into Earth's core.

Snowball Earth: Periods in the history of the Earth when its entire surface was covered in ice, dues to extreme low temperatures.

Soil: The unconsolidated mineral or organic material on the immediate surface of the earth that serves as a natural medium for the growth of land plants.

Source-transport-trap concept: A common framework for the study of mineral deposits, in which the source of the commodity of interest, its transportation, and its immobilization to form a deposit are considered.

Spallation: The process of production of some light elements (most of ^{3}He, and the elements lithium, beryllium, and boron) in the interstellar space, when cosmic rays, mostly fast neutrons, impact interstellar matter.

Spatial geostatistics: The description and prediction of numerical values in space, based on values for sample with known spatial position.

Stable isotope: An isotope that will not change (its number of protons and neutrons will remain invariable) under any natural conditions on Earth.

Standard deviation: In statistics, a common measure of dispersion, defined as the mean of the deviations from the mean.

Stellar nucleosynthesis: The production of new elements, from He to Fe, in the star from the main sequence.

Sub-alkaline: A rock chemically saturated with respect to silica, which has normative minerals such as orthopyroxene and quartz but no nepheline and olivine. Synonymous to tholeiitic.

Subduction: In plate tectonics, the process of one lithospheric plate descending beneath another when the two converge towards each other.

Super-Nova: A powerful and luminous stellar explosion during the last evolutionary stages of a star, when it either collapses to a neutron star or black hole, or it is completely destroyed.

T

Thermochronology: The determination of temperature history of a rock, based on the concept of closure temperature.

Total Alkali-Silica (TAS) diagram: A diagram of $NaO + K_2O$ versus SiO_2, displaying the typical fields of all igneous rocks and used for classification and naming of igneous rocks.

Trace element: An element that is not essential or significant in a mineral or rock and is found in small quantities (typically well below 1 wt%).

Transition metals: The elements scandium, titanium, vanadium, chromium, manganese, iron, cobalt, nickel, and copper, which have a partially filled d subshell of electrons in their atomic structure.

U

Ultramafic: An igneous rock composed mostly of mafic minerals (e.g., hypersthene, augite, or olivine), with overall SiO_2 contents below 45 wt%. Typical examples are dunite, peridotite, and pyroxenite.

Uranium-lead geochronology: An absolute (or isotope) geochronological method, relying on the disintegration of ^{235}U to ^{207}Pb and/or ^{238}U to ^{206}Pb.

V

Variation diagram: Any two-variable diagram for elements or compounds displaying a typical variation for a rock type or a suite of genetically related rocks.

Variogram: In spatial statistics, a function describing the degree of spatial dependence between random variables.

Volcanogenic massive sulphide (VMS) deposit: A type of deposit made of high-concentration copper-zinc sulphides, associated with volcanogenic hydrothermal events in submarine environments.

W

Water treatment residual (WTR): Any solid by-product of water purification for human consumption.

X

X-Ray fluorescence (XRF): A type of X-ray emission spectroscopy in which the characteristic X-ray spectrum of an element in a sample is produced by using X-rays to induce the substance to fluoresce and emit characteristic secondary X-rays.

Index

A

Å 2
Acetate 76
4-acid digestion 76
Acidic 97
Acidic soil 94
Acidity reduction 95
Actinides 3, 4
Additive 95
Adsorbed anion 67
Adsorbed cation 67
Adsorbed element 76
Adsorbed ion 67
Adsorption 67, 75, 96
Aerosols 88
AFM ternary diagram 38, 39
Agricultural pollution 90, 91, 93
Agriculture 90
Alaskite 71
Albedo 88
Albite 52
Alkali-earth metals 4
Alkali metals 4
Alkaline 44, 97
Alkaline group 38
Alkaline series 36
Alluvium 67
Alpha 7
Al saturation 38
Alteration 12, 45, 50, 69
    alteration pattern 51
    clay alteration 50
    qualification 52
    quantification 50, 52
Alteration assemblage 72
Aluminosilicates 13
Alunite 89
Ammonium 92
Amorphous 97
Amphibole 58
Analytical methodology 77
Ancient Greece 87
Andesite 63
Angstrom 2
Anion 3, 67
Anomaly 30, 78
    biogeochemical 67
    definition 30, 63
    statistical definition 31
    surface anomaly 62
Anorthite 47
Anthropogenic lead 91
Anthropogenic pollution 90
Anthroposphere 86, 87, 92, 94
Antiknock 91
Apatite 71
Aquatic life 93
Aqueous fluid 68
Aqueous solution 66, 67
Aquifer 92

$^{40}Ar/^{39}Ar$ geochronology 72
Argillic alteration 69
Arsenic 94, 97
Atmosphere 86, 87, 92, 94
Atomic mass 3
Atomic number 2
Atomic size 2
Atomic spectroscopy 22
    absorption 23
    emission 24
    optical emission spectrometry 23
Australia 39, 52, 53

B

Background 91
Background geochemical heterogeneity 80
Background orientation heterogeneity 82
Basalt
    calc-alkaline basalts (CAB) 40
    geodynamic context 41
    island arc basalts (IAB) 40
    mid-ocean ridge basalts (MORB) 40
    ocean island basalt (OIB) 40
    within-plate basalts (WPB) 40
Base metal deposit 77
Basin evolution 72
Bedrock outcrop 74
Beetle 88
Benthic foraminifera 89
Big Bang 2
Bigeleisen, Jacob 7
Bioaccumulation 68
Bioavailability 94, 96
Biogenic carbonate 89
Biogenic cellulose 89
Biogenic phosphate 89
Biogeochemical prospecting 94
Biogeochemistry 94
Biomass decay 93
Biomass production 95
Biophile elements 86
Biosphere 86, 87, 92, 94
Biotite 58
Bog 66
Bond deficiency 67, 75
Bottled water 93
Bowen Reaction Series 13
Bulk chemical composition 50
Bulk Silicate Earth 39
Bush 67

C

Calc-alkaline group 38
Calcite 52, 75
California brome 95
Canada 52, 58, 68, 72, 77, 91, 93, 95

Carbon 94
Carbonate 67, 70, 94
Carbonate sediment 89
Carbonatite 47, 71
Carbon isotopes 77
Carbon sequestration 94
Carolina poplar 95
Cation 3, 67
Cation composition of rocks 36
Central tendency 29, 30, 78
C1 chondrite 39, 42, 47
Chalcophile element 6
Chalcopyrite 72
Charge density 9
Chart of isotopes 8
Chemical characteristics 6
Chemical dispersion 64, 65
Chemisorption 67
Chert 89
Chile 72, 73
Chloride 71
Chlorite 51, 52, 72, 74
Chuquicamata 72
    Chuquicamata Porphyry Complex 72
Circumpolar region 89
Classification 36
Classification diagram 36, 38
Classification of sedimetary rocks 39
Clay mineral 67, 70
Climate 64, 76, 89
Climate change 87, 94
    astronomical causes 88
    effects on society 87
    forcing feedbacks 88
    fossil fuel 87
    human activity 87, 88
    stabilizing feedback 88
    terrestrial causes 88
Closure temperature 58
Coagulation 96
Colloids 97
Colluvium 64
Colour coding 79
Commodity 62, 86
Common lead 56
    common lead correction 56
Compatibility 47, 63
Compatible element 47
Compost 95
Concordia
    diagram 58
    line 57
Concordia diagram 57
Conservative element 51
Contaminant 97
Contaminated soil 95
Contamination 20
Continental crust 39
Continent-continent collision 40
Contour map 79
Control line 52

Convection cell 78
Coordination number 2, 12
Correlation coefficient 31
Correlation regression 51
Covalent bond 4
Cross-section 93
Crystalline structure 13
Crystal structure 13
Cumulate 47

D

Dacite 52, 63
Data distribution 29
    maximum 30
    minimum 30
    percentile 30
    quartile 30
    standard deviation 30
Data point 79
Data treatment 28
    single variable 29
Daughter isotope 55
    initial incorporation 56
de Montesquieu, Charles 87
Degree of differentiation 44, 48, 63, 89
Delauney polygon 80
δ-notation 6
Deposit 87
Deposit formation
    chemical conditions 70
    fluids 68, 69
    formation timing 72
    physical conditions 69
    pressure 69
    processes 71
    temperature 69
Depositional environment 64
Depth profile 91
Detrital mineral 72
Diagenetic process 72, 89
Diatom 89
Differentiated 62, 71
Differentiation 13, 39, 44, 63
Diffusion 55, 64
    diffusion rate 55
    temperature dependence 55
Dimethylarsenic acid 97
Discordia 57, 58
Discrimination diagram 40, 44, 45
Disintegration constant 9
Dispersion 63, 66, 78
    bioturbation dispersion 64
    erosion 63
    glacial dispersion 64
    weathering dispersion 64
    wind dispersion 64
Dissolution 22, 62, 75
    partial dissolution 22
    total dissolution 22
Dissolved ion 67
Dissolved species 71
Distribution 29
Dolomite 52, 75, 94, 96
Dubos, Jean-Baptiste 87

E

Earth's orbit 88
    eccentricity 88
    obliquity 88
    precession 88
Earth tectonic activity 88
Economic factor 76
Effective ionic radius 2
Eh 12, 66, 71
Electronegativity 3, 10, 14
Electronic configuration 47
Element behaviour 9, 62
    compatibility 9, 13
    mobility 9, 12
    substitution 9, 13
Element bond 3
Element mobility
    in soil 75
Element mobilization 63
Element ratio 40
Elements classification
    by abundance 4
    by reservoir 5, 6
Element substitution 69
    pairs 14
    rules 14
Enriched Mantle (EM) 45
Environment 86
Environmental biogeochemistry 94
Environmental factor 76
Environmental geochemistry 86, 93
Environmental research 86
Enzyme leache 76
Epithermal deposit 70
Erosion 62, 67
Eu anomaly 47, 48
Eutrophication 94
Evaporation 68
Experimental variogram 82
Exploration 50
Exploration geochemistry 93
Extrapolation 79

F

False negative 67
False positive 67
Faraday cage 27
Farming 86
Fawn tall fescue 95
Feldspar 53
Felsic 36, 39, 40, 47
Fe oxide 67
Ferrihydrite 74
Fertilizer 93
Field strength 9
Field technology 77
First melting 71
Fishing 86
Flin Flon-Snow Lake Mineral Belt 77
Fluid flow channel 64
Fluid inclusions 71
Fluid salinity 71

Fluorite 71
Foraminifera 89
Forestry 86
Fractional crystallization 13, 47
Fractionation 89
    basic rule 7
    extent of fractionation 7
    quantification 7
    temperature dependence 7
Fractionation factor 7, 89

G

Galena 70
Gangue 69, 71
Gasoline 90
Gaussian distribution 30
General element ratios 51
Geo-barometer 69
Geobarometry 15
Geochemical anomaly 62, 78
Geochemical classification 4
Geochemical exploration 62, 87
    definition 62
Geochronology 36, 54, 72
    closure temperature 55, 56
    general equation 54
    isochrone method 56
    parent-daughter systems 54
    U-Pb dating 54
Geodynamic environment 40
Geodynamic setting 63
Geometrical weighing factor 80
Geophysical exploration 62
Geophysics 68
Geosphere 86, 87, 92, 94
Geosphere-biosphere interface 94
Geotectonic context 36
Geo-thermometer 69
Geothermometry 15
Germination 96
Gibbsite 75
Glacial deposit 64
Glacial record 88
Glacial till 74
Global air temperature 89
Global Meteoric Water Line 69
Global water cycle 68
Goethite 74
Gold exploration 77
Granite
    ocean ridge granite 40
    syn-collisional granite 40
    volcanic arc granite 40
    within plate granite 40
Granitoid 40, 63
Greenhouse 95
Greenhouse effect 87
Greenhouse gases 87
Greenland 72
Grenville orogeny 58
Groundwater 92, 93, 97
Groundwater exploration 77
Guyana 72
Guy Callendar 87

Gypsum 75, 95

H

Habitat 90
Halite 71
Harding grass 95
Harker diagrams 39–41
Hematite 75
High-μ 46
Histogram 29
Historical temperature compilation 88
Homogenization temperature 71
Hotspot 40
Hume, David 87
Hydrography 76, 77
Hydrolysis 67
Hydrosphere 68, 86, 87, 94
Hydrothermal alteration 48, 50, 62
    quantification 73
Hyper-accumulation 68

I

Ice core 88
Ice-free Earth 90
Igneous activity 72
Igneous rock 36, 63, 71, 89
Illite 50, 72, 74, 96
Immobile ions 12
Incompatible 62, 71
Incompatible element 44
India 77, 93
Indian mustard 95
Inductively-Coupled Plasma (ICP) 23, 24
Industrial activity 90
Industrial pollution 90, 91
Intermediate composition 36, 39, 47
Interpolation by triangulation 80
Interpolation method 81
Inverse-distance interpolation 79, 80
Ion beam 25
Ion exchange 67
Ionic bond 4
Ionic charge 2, 10, 14
Ionic potential 9, 14
Ionic radius 2, 10, 47
Ionic size 14
Isochron 57
Isochron equation 56
Isoconcentration lines 79
Isocon diagram 51
Isopleth 79
Isotope 6, 10
    daughter 9
    definition 6
    in exploration 77
    isotopic ratio 6
    parent 9
    primordial isotope 56
    radioactive 6
    radiogenic 6
    stable 6
Isotope composition 41

Isotope fractionation 69
Isotope system 44
Isotopic composition 6, 93
Isotopic equilibrium 69
Isotopic fractionation 7
    equilibrium 7
Isotopic geochronology 9
Isotopic mixture 69
Isotopic system 44

K

Kaolinite 51, 74, 89, 96
Kriging 79, 81, 82

L

Lag distance 81
Lake 66, 74
Landfill 97
Lanthanides 3, 4, 47
Laser ablation-ICP-MS 71
Last melting 71
Leachant 75
Lead 94
Leaded gasoline 91
Lead isotopes 91
Lead poisoning 90
Lead pollution 90
Leucogranite 47, 71
Ligand 13
Limestone 94, 96
Linear interpolation 79
Linear lateral variability 80
Lines of equal concentration 79, 82
Liquid phase 13
Lithogeochemistry 36, 73
Lithophile 62
Lithophile element 5
Local background 63, 91
^{176}Lu-^{176}Hf 57

M

Macro-nutrient 67
Mafic 13, 36, 39, 40, 47
Magma chamber 13, 47
Magma differentiation 39, 45, 51
Magmatic process 71
Magmatic water 69
Major element 5, 40
Manitoba, Canada 36
Mantle 41
Mantle heterogeneity 41
Mantle reservoir 40
Manure 93
Mass spectrometer 26
    detector 26
    magnetic sector 26
    source 26
Maturity 39
Mean 29, 78
Mean annual rainfall 64

Mean annual temperature 64
Measures of location 29
Mechanical dispersion 63, 67, 76
Median 29, 78
Mercury 94
Metabolic effect 93
Metaluminous 38
Metamorphic water 69
Metamorphism 12
Meteoric water 69, 78
Meteorite 39
Microbeam analyses 27
    electron microprobe 27
    laser ablation 28
    secondary ion mass spectrometry 28
Micro-nutrient 67
Microthermometry 71
Mid-Ocean Ridge Basalts (MORB) 45
Milankovitch cycles 88
Mineral chemistry 69
Mineral deposit 62, 94
    characteristics 68
    evolution 68
    formation 68
Mineral exploration 36, 68
Mineral formation time 56
Mining 86, 90
Minor elements 5
Mississippi Valley-type deposit 70
Mixing calculations 53
    curvature factor 53
    isotopes 53
Mn oxide 67
Mobile cations 12
Mobile ion 67, 75
Mobile Metal Ion (MMI) 76
Mobile oxyanions 12
Mobility 63, 71
Mode 29
Molar proportion 53
Molar ratios 51
Mollusc 89
Montmorillonite 74
Multi-variate analysis 33
Muscovite 52
Muscovite alteration 73

N

Natural environment 86
Natural neighbour interpolation 79, 80
Natural resource 86
^{143}Nd/^{144}Nd ratio 44
Near surface environment 87
Near total dissolution 76
Nitrate 92
Nitrates pollution 93
Nitrogen 92, 94
Nitrogen isotopes 93
Nitrogen pollution 92
Nitrous oxide 92
Noble gases 3, 4
Non-metals 3, 4
Normative mineralogy 36, 50
Noxes 92

Nucleosynthesis 2
    stellar nucleosynthesis 2
Nucleus 2
Nugget effect 81
Nutrition 92

O

Ocean Island Basalts (OIB) 45
Ocean level 89
Ocean water 69, 89
Oil exploration 77
Ontario 58
Oolith 89
Ore 71, 90
Ore-forming process 71
Ore-related alteration 72
Organic additives 96
Organic fertilizer 95
Organic matter 67, 94
Origin of alteration water 69
Orography 67
Ortometamorphic 39
Ostracode 89
Other metals 4
Outcrop visibility 36
Overburden 64
Oxidation-reduction barrier 71
Oxidation-reduction potential 12
Oxidations state 71
Oxidation state 2
Oxidized sulfur 77
Oxidizing 71
Oxidizing environment 13
Oxyanion 67
Oxygen 94
Oxygen isotopes 40, 77, 78, 93
    in sedimentary record 90

P

Paleo-climate variations 89
Paleo-temperatures 88
Partial dissolution 75
Partial leaching 75
Partial lead loss 57
Partial melting 13, 41
Particle size distribution 97
Parts per billion 5
Parts per million 5
Past climate change 88
Pathfinder element 63
Pauling's electronegativity scale 4
Pb isotopes 44
$^{207}Pb/^{206}Pb$ age 57
$^{207}Pb/^{206}Pb$ age equation 57
$^{206}Pb/^{207}Pb$ ratio 91
Pearce element ratios 51, 73
Pearson product-moment correlation coeffi-
    cient 32
Pegmatite 47, 71
Peralkaline 38
Peraluminous 38
Periodic table 2–5, 86

Pesticide 97
PH 66, 67, 75, 94, 96, 97
    near neutral 67
Phase transition 71
Phosphorus 97
Photosynthesis 86
Physisorption 67
Phytorestoration 95, 96
Plant ingestion 97
Plant species 68
Plasma torch 23
Point estimation 79
Pollen 88
Pollutant 87, 90, 91
Pollutant dispersion 94
Pollutant transportation 91
Pollution 86
Pollution remediation 96
Pollution source 91
Pollution transport mechanism 91
Polymerization 67
Polymers 97
Polymetallic deposit 62
Porphyry copper 73
Porphyry copper deposit 63, 70, 72
Portable XRF 26
Potassic alteration 69, 70, 72
Potassic feldspar 58
Prairies 74
Precipitation 68, 97
Primary mineral 66
Principal component analysis 33
Prospecting 62, 68
Proton 2
Pyrite 69, 72
Pyrrhotite 69

Q

Quadrupole 25
Quadrupole mass spectrometer 24, 76
Quartz monzonite 70
Quartz-sericite alteration 72

R

Radioactive decay 9
Radioactive disintegration 9
Radio-frequency 23, 25
Radiogenic 56
Radiogenic isotope loss 56
Radiogenic isotopes 48
Radiogenic lead 57
Rare Earth Elements 47
Rate of decay 9
$^{87}Rb-^{87}Sr$ 57
Reduced sulfur 77
Reducing 71
Reducing environment 12
REE pattern 47, 49
REE spectrum 45
Remediation 86, 91, 94, 96
Re-Os geochronology 72
Representative samples 18

Residual overburden 64
Residual uncertainty 81
Reubens Canada bluegrass 95
Rhyodacite 63
Rhyolite 52
Rice 97
Rifting 40
Rocks chemical variability 39
Root system 68
Russia 68
Rutherford 9
Ryegrass 95

S

Sample preparation 21
Sample size 76
Sample treatment 21
Sampling 18
    of vegetation 20, 76
    quality control 18
    representativity 18
    scheme 18
    size of sample 19
    soil sampling 20, 73
    spatial resolution 77
    stream sediments 76
    surface water 76
    taking care of samples 20
Sampling budget 77
Sampling medium 74, 76
Sampling pattern 76
Sampling program 77
Sandstone basin 50
Sandstone classification diagram 39
Saturnism 90
Scanning electron microscope 97
Sceptic 93
Secondary electron multiplier 25, 27
Secondary mineral 66
Sediment 97
Sedimentary record 90
Seed germination 95
Selenium 94
Semi-metals 3, 4
Semi-variance 81
Septic 93
Sericite alteration 69
Siderite 52
Siderophile element 5
Silicate melt 47
Silicates 13
Size-coding 79
Skewed distribution 30
Smectite 89
Smelter 91, 94
Smelting 90
$^{147}Sm-^{143}Nd$ 57
Snowball Earth 88, 90
Sodium arsenate 97
Soil 67, 74, 94, 96
    classification 74
    development 74
    horizons 74
    mineralogical composition 74

types 74
variability 74
Soil chemistry 94
Soil pollution 94, 96
Soil profile 20
Soil remediation 95, 97
Soil sampling 75
clay extraction 75
contamination 75
sample preparation 75
Solid phase 13
Solution 66
Sorption 67, 97
Source-transport-trap concept 62
Spallation 2
Spatial data visualisation 79
Spatial geostatistics 78
Spearman rank correlation 32
Speciation 67
Sphalerite 69
$^{87}Sr/^{86}Sr$ 89
$^{87}Sr/^{86}Sr$ ratio 44, 89
Stable isotope 69
Statistical treatment 78
Streaker redtop 95
Stream 66, 74
Stream sediment 74
Stromatolite 89
Structural charge 67, 75
Structural formula 51
Sub-alkaline group 38
Subalkaline series 36
Subaluminous 38
Subduction 40, 63
Substitutions 13
Sulfide ore 91
Sulfur isotopes 77
Sulfur isotopic composition 78
Sulphide mineral 77
Summary statistics 78
Super-Nova 2
Surface anomaly 66, 78, 79, 90, 96
detection 72
Surface bonding site 67
Surface geochemical data 79
Surface media 90, 91
Surface sampling media 91
Surface water 74
Survey design 76
Sustainability 86
Svante Arrhenius 87
Symmetrical distribution 29

T

Technosphere 86
Tectonic provenance 40
Tectonic setting 40
Tectonic setting of sediments 41
Thermochronology 58
Thick section 71
Thiessen polygon 80
Tholeiitic 44
Tholeiitic series 36
Thomas Jefferson 87
Thorium 71
Th/U ratio 71
Topography 64, 76, 79
Total Alkali vs. Silica (TAS) diagram 36, 37
Total dissolution 36
Toxic environment 94
Toxic metals 94
Trace element 5, 40
Transition metals 3, 4
Transport 62
Trap 62
surface trap 62
Trapping pressure 71
Tree 67, 68, 76
Tree health 76
Tree ring 88
Tree species 76
Treetop sampling 76
Tributary 76
Turbidite 52
Two variables 31
Tyndall, John 87

U

Ultramafic 36, 39, 47, 62
U-Pb geochronology 72
Upper mantle 39
Uraninite 71
Uranium deposit 71, 72
Uranyl complex 71
USA 52, 70, 78, 94, 97

V

Valance 2, 47
Valance electron shell 2
Valence electron 4

Value 45
Van der Waals force 67
Vapour feedback 88
Variance 81
Variation diagram 38, 39
Variogram 81
experimental variogram 81
model variogram 81
range 81
Vegetation 67, 74, 94
Vein gold deposits 69
Vermiculite 74
Visual representation 31
VMS deposit 52, 63, 77
Volatile 41, 63
Volcanic activity 88
Volcanic rocks 36
Volumetric percentage 51

W

Waste 86
Water reservoir 92
Water/rock ratio 69
Water treatment 97
Water treatment plant 96
Water Treatment Residuals (WTR) 96, 97
Weather 67
Weathered profile 64
Weathering 12, 88
Weighing 80
Weighing factor 80, 81
Whole rock classification 36
Whole rock geochemistry 36, 72

X

X-ray fluorescence 26

Z

Zero-distance variability 80, 81
Zircon 56–58

Printed in the United States
by Baker & Taylor Publisher Services